# MUSHROOMS FOR ALL SEASONS

## Part 4
## Winter & Beyond

# RICHARD WINDER

This is not a field identification guide.

Mushrooms have poisonous look-alikes and can cause allergic reactions or other illness in some individuals. Incorrect identification, improper harvest or storage, and unsafe cooking methods can result in illness or injury. It is the reader's responsibility to obtain expert guidance and to correctly identify, harvest, store, and cook their mushrooms.

© 2024 Richard S. Winder
ISBN: 978-1-0688613-3-8

Cover design and art by R. Winder
Interior layout and design by R. Winder

Richard S. Winder
5614 Woodlands Rd., Sooke, B.C. V9Z 0G5 CANADA

# Other books in this series:

Available at: https://www.lulu.com/spotlight/rswinder

**(Part 1)** *Beginnings & Springtime*: This book has an introductory section for cooking with mushrooms, covering basic methods, sauces, etc. It also has sections for morels, oyster mushrooms, fairy ring mushrooms, and shaggy manes.

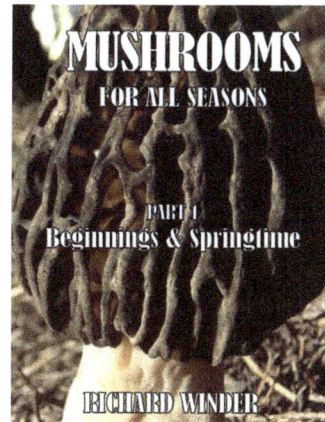

**(Part 2)** *Summer*: This book has sections for *Agaricus* spp. (including button mushrooms, portabellas and creminis), chanterelles, summer truffles (including *Tuber aestivum*), wine caps, puffballs, bamboo mushroom, paddy straw mushroom, hen-of-the-woods (maitake), black wood ear, and lobster mushroom.

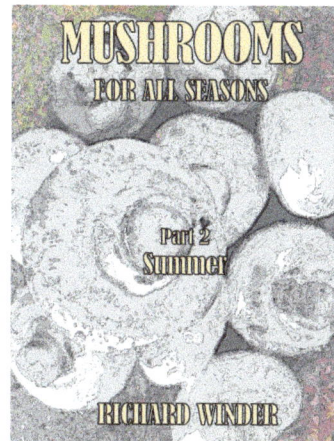

**(Part 3)** *Autumn*: This book has sections for boletes, saffron milk cap, cauliflower mushroom, lion's mane, shiitake, hyp pop (shimeji), pine mushroom (matsutake), and fall truffles (including North American truffles and Italian white truffles).

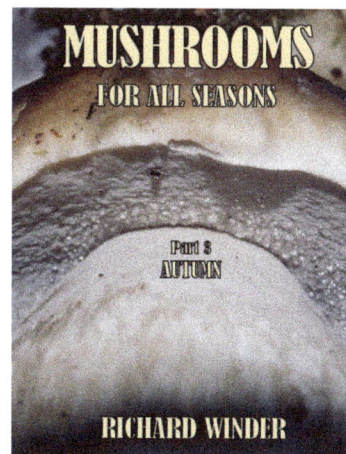

Dedicated to the memory of Burton Taylor
Bakery owner, violinist, dynamite hauler, volunteer fireman, and all-around great grandpa

# Foreword

Fungi are fascinating creatures with an array of abilities, some of which we share and others that would horrify us if we considered doing them ourselves. Imagine, for instance, that you sense the presence of a delectable dessert, grow your fingers over to it, emit enzymes from the tips of your fingers and absorb liquefied pie back through your fingertip skin. Arrgh! Although we animals and fungi share the heterotrophic lifestyle (using plants, animals, mushrooms, and algae as food), our methods of acquiring that food differ; fungi absorb food, while we ingest it first. Many fungi break down dead organic matter such as tree branches and logs; some of these wood decomposers are cultivated for food, e.g. Oyster Mushrooms, or used to create innovative products to replace animal leather and petroleum-based Styrofoam™. Some fungi cause diseases in living hosts, such as Athlete's Foot, or the white dusting called Powdery Mildew on squash leaves in your garden. But the fungal lifestyle relevant to truffles is the mutually beneficial symbiosis known as mycorrhiza in which the plant host provides the fungus with sugar from photosynthesis and the fungus provides nutrients like nitrogen and phosphorus from the soil to the plant roots. Many of our choice edible forest mushrooms, such as boletes or the chanterelles and hedgehogs mentioned in this book, are the fruiting bodies of mycorrhizal fungi—as are truffles.

Truffles are best known and appreciated for their rich and complex aromas. During its maturation process, the Italian white truffle has been demonstrated to produce over 200 volatile aromatic compounds. The purpose of these is to attract animals to find the truffle, dig it up, eat it and disperse the spores away from the maternal mycelium to a new site where the animal defecates, thereby delivering spores to the surface of the ground. Truffle spores that travel through a digestive tract are primed to germinate and grow in search of a new host plant root to colonize. Truffles, therefore, are part of a three-way co-evolved symbiosis: truffle fungus—plant host—animal disperser.

Certain of the most valued culinary truffles can be cultivated in orchards with living plant hosts such as oaks and hazelnut trees. Truffles that are native to Europe are now cultivated around the world where soil and climate are suitable. Although we know that our west coast native choice culinary truffles grow with Douglas-fir, despite years of effort these native truffles are not yet produced under cultivation. The truffle orchards in BC that are producing cultivated European truffles (Burgundy truffle, *Tuber aestivum*, and Bianchetto truffle, *Tuber borchii*) have web sites so purchases from them can be made in season. Truffle production under cultivation and management, including irrigation, can also vary tremendously from year to year. And truffle production in orchard does not last forever, with evidence suggesting that production may decline after 10-15 years. The orchard in BC that has been producing black Périgord truffles (*Tuber melanosporum*) for about a decade is now experiencing a decline in production as has also been reported from Europe. Truffle cultivation is a risky and costly enterprise and those who are willing to give it a try are to be admired for their courage, determination and vision!

Although female pigs were once used in Europe to hunt for truffles, most truffles (both cultivated and wild) are now found by trained truffle dogs. Some people have harvested truffles using rakes, but this technique is inferior because trained truffle dogs find ripe truffles with mature aromatic profiles and optimal value whereas raking is non-selective and most truffles found are immature, poor in aroma and low in value. There is also concern that raking the forest floor is destructive to both plant roots and fungal mycelia. Dogs can easily be trained to find truffles using the same scent-training approach used to train them to find drugs, explosives, cadavers, and endangered species. The handlers of truffle dogs in BC tell me that it takes a great deal of searching to find native culinary truffles and that year-to-year changes in weather and resultant truffle production can be quite large, leading to lots of time searching with little success even at sites where truffles have been found in previous years.

As Richard has described, truffle products come in various degrees of naturalness, with most lower-value foods like chips flavoured by food chemists using the most common volatile organic compounds produced by truffles, including 2,4-dithiapentane (bis-(methylthio)-methane, dimethyl sulphide, and 2-methyl-butanal. A recent Spanish study demonstrated that study participants with no experience of real truffles rated artificially flavoured products much higher than did participants experienced with real truffles. The inexperienced group also associated negative attributes (weird and disappointing) to products made with real black truffle, whereas experienced truffle consumers associated positive attributes (truffle flavor, truffle smell and gourmet) to them. Bottom line, the aromas and flavours of actual truffles are quite different from those of artificially produced flavours.

My experience infusing with truffles is almost entirely based on Oregon white truffles purchased or gifted at the Oregon Truffle Festival. I put my five or six little truffles into an open paper bag at one end of a large plastic food storage container and the foods to be infused beside but absolutely not touching them—butter, eggs in shell, and nuts. Within a day of infusing in the fridge, the foods had powerfully absorbed the truffle aroma. Usually, I could then replace the first set of foods with another and infuse for another day. At the end, I could then use the actual truffles by grating them onto scrambled eggs or cooked pasta. Prominent in my memory though, is the well-marbled beef that I infused and then used to make Beef Stroganoff, inspired by a winery lunch that was part of the Oregon Truffle Festival. Superb!

Buying truffles, especially local truffles, can be challenging and costly. In British Columbia, our excellent native truffles, the Oregon or Western White Truffles (*Tuber gibbosum* and *Tuber oregonense*) and the Oregon or Western Black Truffle (*Leucangium carthusianum*), can be harvested from Douglas fir forests with the help of a trained truffle dog but currently I am not aware of anyone in British Columbia marketing locally harvested native truffles.

Perhaps one day, when your budget permits, you will get to experience fully mature fresh truffles but in the meantime, enjoy the truffle essence highlighted in Richard's wonderful, warming, wintertime recipes.

Shannon Berch PhD
Retired Research Scientist
Co-founder of South Vancouver Island Mycological Society and Truffle Association of BC

# Table of Contents

# Introduction

One dark and stormy night in December, 1993, six people got together in Victoria, British Columbia (B.C.), Canada to form the South Vancouver Island Mycological Society (SVIMS). Yours truly was one of those founding members. As the club moved forward through the years it thrived and gained many new members from the region and some from beyond. It undertook a campaign of informing mycophiles (mushroom-loving people) about identification, biology, ecology, and other aspects of mushroom lore. Forays into the woods were held, a newsletter was published, and yes, 'survivors' banquets were well-attended (some mycological humour, there) with great food and recipes shared by all. In B.C. and the Pacific Northwest, a different cohort of mushroom species emerges every season and brings a chance to experience them anew. It's always an adventure, even if it's only to get some fresh air, take a hike into the forest, and observe them first-hand. This book is part of a series that attempts to capture that blended spirit of curiosity, learning and enjoyment, and share it with a wider audience.

You'll notice that there are quite a few cautionary caveats about consumption in this work. In fact, if you choose to ignore them, you assume responsibility for the results. However, the intent here isn't to scare people away from consuming mushrooms. No, it's hoped that encouraging people to be safe will allow them to enjoy choice edibles with greater confidence. So, please understand that the illustrations and text in this book in no way offer the information you will need to safely and reliably identify wild-harvested mushrooms—**this is not a field guide**. To quote one venerable source, "Reference books are not a talisman (and the illustrations in this book are decorative—not for identification purposes)."[1] So, while the choice edibles in this book can sometimes be wild-harvested, you should be fully aware that there are some poisonous and even a few deadly species, and that you need to take precautions to avoid them. Obtain your mushrooms from safe and reliable sources, or if you want to harvest them in the wild, you should learn how to identify them from experienced people using reliable and up-to-date resources. Rather than trying to cram as much advice as possible about identifying mushrooms into an already large cookbook, this work instead opts to summarize interesting ecological, nutritional, and cooking aspects of the more popular species. In this way, learning about identification can focus on sources more thoroughly dedicated to that task. The resource section at the end of this book will hopefully lead you to one of the many regional groups, field guides, and texts that can help you learn how to properly identify mushrooms.

As you will see in the nutritional information included for each mushroom, there are always factors to consider that have a bearing on safe levels of consumption. Although once again confusion with poisonous species is always a particular concern, edible mushroom species actually contribute to the bulk of poisoning reports. The reasons can be broken down to:

- Long-term storage of fresh wild-harvested mushrooms without appropriate treatment
- Improper cooking
- Inappropriate long-term storage of prepared mushrooms
- Pest infestations
- Consumption of certain species with alcohol
- Chemical and mineral contaminants
- Frequent consumption (morels in particular)
- Deliberate use of hallucinogenic fungi.[2]

Allergies or 'idiosyncratic' (rare or unique) reactions are sometimes encountered and should not be dismissed out-of-hand. Before you panic, remember that these factors are also the case for other foods in the plant and animal world. The author reliably experiences (for a brief period) a generalized sensation including ringing in the ears, slight dizziness, and slightly lower blood pressure after consuming various species of wild-harvested mushrooms. While that makes him cautious about what he eats and how much, it hasn't stopped him from enjoying mushrooms and writing about them!

---

1    Grigson, J. 1975.
2    Govorushko, S., *et al.* 2019.

The key thing here is to follow the guidelines and to know your limits. Keeping a few pointers in mind can help you to avoid disappointments and increase enjoyment. In his wonderful field guide *Mushrooms Demystified*,[1] David Arora offers some advice about wild mushrooms and mushroom cooking that can be paraphrased and 'boiled down' to the following:

- Avoid uncertain identification or condition—When in doubt, throw it out."
- Another thing that should go without saying: don't eat anything rotten, including mushrooms.[2]
- Moderate. Don't over-stuff yourself just because there's a lot—digestive upsets may follow.
- Trying a new one? Just eat a little (once cooked) and let a reasonable time pass to be certain you won't react.[3]
- Don't serve large groups a mushroom they've never tried before.
- Don't mix with species you've never eaten—if there's a reaction, it will be harder to sort out.
- Yet another thing that really should go without saying: avoid mushrooms with lots of maggots.
- Try to use them fresh—flavour can degrade during storage.[4]
- Mushrooms are insipid after a heavy rainfall unless dried to concentrate flavor.
- Don't overdo spices, oils, wine, etc.—mushroom flavours are subtle and easily overpowered. Seek balance.
- Give choice edibles a few chances—lots of variables can influence a dish and change your opinion.

In addition to the above, avoid mushrooms from contaminated roadsides, chemical-treated lawns and burn sites, mining sites, etc. Certain mushrooms such as matsutake should be avoided if they have been exposed to frost. If wild-harvested when they are too old, mushrooms can also cause problems (lobster mushrooms are notorious for this).

This book is the fourth of a four-part series that focuses on the most popular and widely available choice edibles for those living in B.C. and the Pacific Northwest (both cultivated and wild-harvested, locally or abroad), categorized by season. This book focuses, broadly, on those that may fruit during the winter season (Fig. 1). At the end, it also includes information on some other ingredients that feature quite often in mushroom recipes. Regarding season, some mushrooms actually fruit in multiple seasons or can be purchased year-round, so there's some hand-waving involved when it comes to covering the seasons. Also, some mushrooms that are popularly cultivated or consumed are not included. That's not necessarily a deliberate judgment on traditional usage or popularity, but simply a focus on the safest options. For example Nameko mushrooms (*Pholiota nameko*) are popular in Japanese cuisine.[5] Elsewhere, chestnut mushroom (*P. adiposa*) is cultivated for consumption[6] but some sources aver that it should definitely not be considered edible.[7] There is one recent case of illness happening after repeated consumption of the cultivated form in North America.[8] On top of that, there are other *Pholiota* species in North America that are either poisonous or with controversial claims of edibility (e.g. *P. squarrosa* and *P. squarrosoides*).[9] For all of these reasons, *Pholiota* species aren't included here. Similarly, other groups with significant controversies, knowledge gaps, lack of availability, limited occurrence, or very cryptic identification features are skipped, trusting that it will eventually all be sorted out in future references.

This series is more than just a set of recipe books. Articles for the various mushrooms in each season go into some detail about their history, biology, ecology, harvest and cultivation, and nutritional characteristics (summarized in Table 2). The series consists of four parts: *Beginnings & Springtime* (Part 1), *Summer* (Part 2), *Autumn* (Part 3), and *Winter & Beyond* (Part 4). Even individual recipes provide some history and background, where appropriate. Part 1 also begins with a section on the basics of cooking with mushrooms and general methods that apply to any season. The brief footnotes in each section refer to full citations in the bibliography at the end of the book.

Part 1 of the series covers the methods for making some of the ingredients common to many of the recipes in this book, for example duxelles, mushroom broth, mushroom ketchup, and seasoned mushroom powder. You should also consult it for best results. Duxelles are a great way to temporarily store mushrooms, and they are versatile in the ways that they can be used in cooking. Most recipes calling for vegetable or meat stocks could just as easily include

---

1     Arora, D. 1986.
2     Even commercial mushrooms are sometimes recalled for contamination with *Listeria*, *E coli*, etc. If 'choice' mushrooms aren't wholesome, they aren't choice.
3     What amount of time is reasonable? Some recommend 24 hours. But some mushrooms have toxins that take days or weeks to manifest. Be aware of look-alikes for your particular mushroom, their properties, and govern yourself accordingly. Best practice is to gain the advice of an experienced person.
4     That being said, drying certain species (*B. edulis*, *Morchella* spp., *L. edodes*) is thought to intensify and/or improve flavours.
5     Pegler, D.N. 2003.
6     Shimizu, K., et al. 2003.
7     First Nature. 2021.
8     Beug, M. 2020.
9     MacKinnon, A. and Luther, K. 2021. (see p. 268 *P. limonella* group)

mushroom broth instead. And finally, you'll be wanting the mushroom ketchup and mushroom powder recipes for a quite a few of the other recipes throughout the book—a good place to roll up your sleeves, put on your apron, and get started!

With all of the above in mind, the recipe sections for each mushroom will start with simple, popular, or traditional ways to prepare them that highlight their unique flavor. But there are others that use the mushroom as a complimentary component. While the emphasis is on those mushrooms and foods available in the B.C., Canada, or North America in general (including grocery stores, health food stores, farmer's markets, gardens, marketing websites, import specialists, cultivation kits, etc.), dishes from around the world make an appearance—because mushrooms are truly part of a global cuisine. There are also a few recipes that are a little fancy. Just keep in mind that the author is a peasant cook—he's a mycologist, yes, but he doesn't also pretend to be a highly trained chef. That means that there's no assumption here that you, dear reader, need a degree in culinary arts to prepare and enjoy the recipes. Many of the recipes have been adjusted to be accessible to the average person and to encourage experimentation. It's true that different mushrooms have different flavours that can be fully appreciated in particular ways, but in general they are also very versatile and can be used to complement a wide variety of dishes and cooking styles. Dozens upon dozens (if not hundreds) of cookbooks stuffed with mushroom recipes can't be wrong! Cooking with mushrooms need not involve much more than a few minutes of happiness in the kitchen—although it can be fun to learn something new, too. And you know, a bolt of lightning from the mushroom gods isn't going to strike you if you fancy changing something in the recipes to better suit your tastes, or want to take a recipe to that next level. Isn't that part of the fun? All of the recipes were personally tested by the author and his family—they were included if their preparation was relatively straightforward, somewhat flexible, and the dish passed the 'taste test.'

Above all, there's one final piece of advice to be added here: Enjoy!

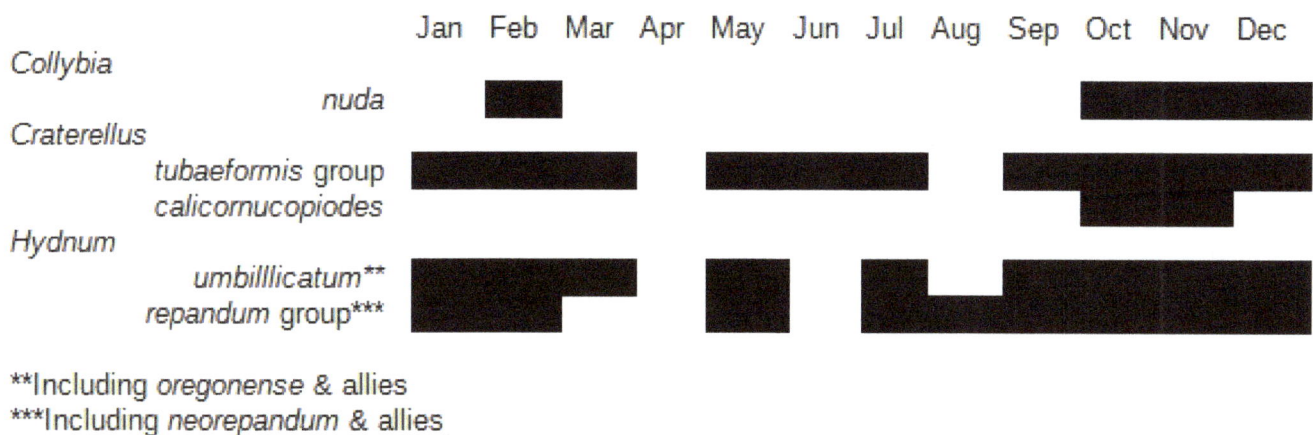

Figure 1. A list of winter mushrooms reported from southern Vancouver Island showing when they were found. Data is taken from the South Vancouver Island Mycological Society Checklist.[1]

---

1     South Vancouver Island Mycological Society. 2022.

Table 1. Relative nutritional content of some choice edible winter mushrooms per 100 g.

| | C. nuda (Blewit) (Dry weight)[1] | C. cornucopiodes (Black trumpet) (? or dry weight)[2] | F. filiformis (Enokitake) (Fresh weight)[3] | F. filiformis (Enokitake) (Dry weight)[4] | H. repandum (Hedgehog) (Dry weight)[5] |
|---|---|---|---|---|---|
| Energy (kcal) | 334-367 | 388?-1583 | 37 | 316 | 1615 |
| Protein (g) | 8-12 | 50?-69 | 2.7 | 23 | 15 |
| Fat (g) | 1-2 | 5-6? | 0.3 | 9 | 8 |
| Carbohydrate (g) | 73-77 | 13-34? | 7.8 | 67 | 62 |
| Fibre (g) | | | 2.7 | 23 | |
| Sugars (g) | | 10.8 | 0.2 | 2 | |
| Calcium (mg) | | 4? | 0 | 0 | 8 |
| Iron (mg) | 0.08 | 8? | 1.2 | 10 | 0.02 |
| Magnesium (mg) | 1 | 50? | 16 | 137 | 324 |
| Phosphorus (mg) | | | 105 | 897 | |
| Potassium (mg) | | | 359 | 3068 | 2805 |
| Sodium (mg) | | 62? | 3 | 26 | 537 |
| Zinc (mg) | 0.1 | 2? | 0.7 | 6 | 0.4 |
| Copper (mg) | 0.08 | | 0.1 | 1 | 0.3 |
| Manganese (mg) | 0.03 | 8? | 0.1 | 0.6 | |
| Selenium (µg) | | | 2.2 | 18 | |
| Vitamin C (mg) | 10 | | 0 | 0 | |
| Thiamine (mg) | | | 0.23 | 2 | |
| Riboflavin (mg) | | | 0.2 | 2 | |
| Niacin (mg) | | | 7 | 60 | |
| Pantothenic acid (mg) | | | 0.2 | 12 | |
| Vitamin B6 (mg) | | | 0.1 | 1 | |
| Folate (total) (µg) | | | 48 | | |
| Choline (µg) | | | 47.7 | | |
| Betaine (µg) | | | 1.4 | 410 | |
| Vitamin E (µg) | | 187 | 0.01 | 0.01 | |
| Vitamin D2+D3 (IU) | | | 5 | | |
| Vitamin D2+D3 (µg) | | | 0.1 | 1.7 | |
| Fatty acids, sat. (g) | | | 0.03 | 0.2 | |
| Fatty acid, unsat. (µg) | | | 0.12 | 1 | |
| Cholesterols (µg) | | | 1 | 4.2 | |

1    De, J., et al. 2022.
2    Barros, L. 2008. ? indicates fresh or dry weight as reported in Ozuna-Valencia, K. et al. 2023.
3    According to United States Department of Agriculture, Agricultural Research Service. 2019 (Proportion of fresh weight) and Stamets, P. 2005. (Proportion of estimated fresh weight)
4    United States Department of Agriculture, Agricultural Research Service. 2022
5    Jedidi, I., et al. 2017.

# Trace elements

Table 3 summarizes trace elements reported from the various winter-fruiting mushrooms in this book (each mushroom's chapter may contain further information regarding mineral or trace elements and contaminant uptake). Fungi often accumulate various elements and compounds in their fruiting bodies. This 'bioaccumulation' happens as the mycelium grows and absorbs nutrients out in the environment, and some of those nutrients can be transported (along with other elements and compounds) through mycelial networks to primordial fruiting bodies as they form and grow. This is a particularly interesting phenomenon in mycorrhizal fungi that are symbiotic with plants. In mycorrhizal relationships, we often focus on the uptake of minerals for the plant in exchange for nutrients for the fungus. But things like heavy metal trace elements that could be detrimental to plant growth are also scavenged from the soil and moved to the fruiting bodies, another potential net benefit for the plant.[1] Implications for human health are probably a case-by-case consideration, relative to different species, different geographic populations and sub-types, and different locations having different soils with different concentrations of trace elements or contaminants (for example areas with high human disturbance such as mining and industrial sites, areas treated with pesticides and other sprays, paved roadsides, etc.). Also, keep in mind that there is a difference between having the potential to accumulate a trace element or contaminant in fruiting bodies, and proving that it is bio-available or actually present in a given locale. There are too many variables to discuss here, but again summary information is provided to better inform the reader. The author treats mushrooms (particularly wild-harvested mycorrhizal ones) a bit like liver—full of vitamins and nice to have once in a while, but perhaps not something to have at every meal.

Table 2. Abundant and/or accumulated trace elements in some edible winter mushroom species. Levels are indicated as relatively abundant or unsafe especially in polluted areas (+) or accumulated but normally safe (~).

| Species | Element[2] | | | | |
|---|---|---|---|---|---|
| | As | Cd | Cs | Hg | Pb |
| *C. nuda* (Blewit) | ~ | | | + | + |
| *C. cornucopiodes* (Black trumpet) | | | + | | |
| *C. tubaeformis* (Winter chanterelle) | | | + | | |
| *F. filiformis* (Enokitake) | + | | | + | + |

---

1   Colpaert, J.V., *et al.* 2011.
2   As = Arsenic, Cd = Cadmium, Cr = Chromium, Cs = Cesium, Hg = Mercury, Pb = Lead. A plus (+) indicates levels that are relatively abundant or unsafe especially in polluted areas .

# Bioactivity

Table 4 summarizes various types of bioactivity and potential human health impacts for the winter-fruiting mushrooms covered in this book. Regarding the potential medicinal or pharmaceutical properties of any particular fungus, please be aware that this work takes no definite stance of advocacy for or against. For an extended discussion and balanced perspective regarding bioactivity and properties of mushrooms affecting health, please consult Part 1 of this book series. There are references that go into this topic in depth,[1] and the reader is encouraged to explore them. Reports are summarized here to better inform the reader, but there is no intent to advocate for or against treatments based on that information, establish precise levels of risk, or suggest recipes for medicinal use. Consult a responsible medical expert for those things!

Table 3. A summary of bioactivity reported in studies of some edible mushroom species fruiting in the winter. Activity is indicated as positive or present (+), unremarkable (~), uncertain (?), or absent (-) for the particular activity. **Indication of activity in this table is not a guarantee of health effects**, merely a listing of what has been reported. Consult medical professionals for further information. For citations leading to this information, please see the Bioactivity section in the relevant chapter of this book.

| | C. nuda | C. tubaeformis | C. cornucop-iodes | F. filiformis | H. repandum | T. indicum | T. melanosporum |
| --- | --- | --- | --- | --- | --- | --- | --- |
| | Blewit | Yellowlegs | B. trumpet | Enoki | Hedgehog | Asian blk. truffle | Blk. Périgord truffle |
| Immunomodulation | + | | + | + | + | | ? |
| Anti-cancer (tumour) | + | | + | + | | | |
| Anti-cancer (other) | + | | | + | | | |
| Mutagenic/genotoxic | | | | | + | | |
| Anti-mutagenic | + | | | | + | | |
| Antibiotic (bacteria) | + | | + | | + | | + |
| Antibiotic (fungi) | + | | + | | + | | |
| Antiviral | | | | + | | | |
| Insecticidal | + | | | | | | |
| Anti-oxidant /anti-aging | + | | + | | + | + | + |
| Anti-inflammatory | | + | + | + | + | | |
| Anti-asthmatic / anti-allergic | | | + | + | | | |
| Hypolipidemic (anti-arteriosclerosis) | + | | + | | | | |
| Anti-diabetic | | | + | | | | |
| Hypotensive / anti-hypertensive | + | | + | + | | | |
| Anti-cytotoxic / organ protective | | | + | | + | | |
| Cytotoxic | | | + | + | + | | |
| Aphrodisiac/anti-impotence | | | | | | | ? |
| Hormonal /estrogen mimic | | ~ | | | | | |
| Creatine kinase (muscle loss) | | | | + | | | |

---

1    Benjamin, D. 1995.

# Winter

# Blewits

Blewits in a basket.[1]

---

1    This is not a field guide.

# Biology & Ecology

**Are blewits really winter mushrooms?:** In southern Vancouver Island, blewits tend to appear with the onset of sharper, colder weather in the latter part of October, and they will fruit right up until a heavy frost, perhaps on into early December if it is relatively mild. In that context they are more like a harbinger of colder weather to come. So, while the official winter season begins with the solstice later in December, blewits are like a harbinger of the colder season that is quickly approaching. We'll therefore park them in this part of the series (Part 4) to keep the other wintertime mushrooms company.

**Fun and names:** When it comes to having a scientific name, the wood blewit has had quite a journey. Well, to be more accurate, the taxonomists studying it have had quite a journey—the mushroom hasn't actually gone anywhere. The mushroom started its academic life in 1790 when Pierre Bulliard named it *Agaricus nudus*.[1] That worked well enough until 1871, when two different mycologists placed it in different genera: *Tricholoma* and *Lepista*. It resided in either genus until 1969, when H.E. Bigelow and A.H. Smith reclassified *Lepista* as a subgenus of *Clitocybe*. But wait! DNA studies uncovered some troubling problems, as DNA studies often do. It seemed that *Lepista* and *Collybia* species were closely related to the core group that formed *Clitocybe*, and that they were polyphyletic (possessing multiple lines of ancestry). Basically, *Lepista* wasn't a coherent genus from the standpoint of phylogeny. *Lepista*, *Clitocybe*, and *Collybia* all seemed to include species more closely related to members of each other's grouping. Even worse, it seemed that the wood blewit wasn't closely related to the type species (ideal example) for *Lepista* (*L. densifolia*), and samples of wood blewit from Asia appeared to be polyphyletic (comprising different species).[2] Until recently, this led to an uneasy situation where wood blewits were called either *Clitocybe nuda* or *Lepista nuda*, depending on the authority. Fast forward to the present and a more extensive look at DNA in these mushrooms has thrown all of that out the window. After further taxonomic renovations and rearrangements, the name proposed for wood blewits is *Collybia nuda*.[3] As this is being written, experts and academic sources are still in the throes of making the switch. Not wanting to seem out of fashion, this work will accord to this latest convention.

Other common names for *C. nuda* in English include blue hat, blue cap, blue leg, or blue stalk mushroom. In other languages, names include *pied bleu* ('blue foot,' French), *ryadovka fioletovaya* (Russian), *zǐ dīngxiāng mó* ('lilac mushroom,' Chinese), *Violetter Rötelritterling* (violet-red knightling, German), *blå ridderhatt* ('blue knight's hat, Norwegian), and *murasakishimeji* ('purple shimeji,' Japanese). Species related to *C. nuda* and also having the 'blewit' moniker include *Collybia sordida* (= *L. sordida*, a.k.a. sordid blewit, dirty blewit, fairy ring fungus, or flesh-purple blewit), *Collybia personata* (= *L. saeva*, *C. saeva*, field blewit, purple-stemmed wood blewit), and *Lepista glaucocana* (a.k.a. *Clitocybe glaucocana* or lavender-coloured wood blewit, according to some a pale variety of *C. nuda*), and *L. panaeolus* (a.k.a. gray wood blewit, spotted blewit). *C. sordida*, *C. personata*, and *L. glaucocana* occur in both Europe and North America. *C. sordida* is a smaller, less robust mushroom that nevertheless closely resembles *C. nuda* and confusingly occurs in the same type of habitat. *C. personata* favours grasslands more than forests (although it can be found in forests, too). Both species are reported to occur in B.C.

**Blewit vs. Bluet:** In the interest of clearing up some confusion, there are names that sound like 'blewit' for lots of different things. 'Bluet' can refer to plants in the bedstraw (a.k.a. madder) family (a group also called 'innocence' or 'Quaker ladies'). These include the Azure bluet (*Houstonia caerulea*), long-leaved bluet (*H. longifolia*), and small or tiny bluet (*H. pusilla*). Beyond the bedstraw family, there is also *Centaurea montana* (a.ka. mountain bluet, mountain cornflower, mountain blue knapweed, perennial cornflower, or perennial bachelor's button). In Québec (but not France), *bleuet* refers to blueberries (hence you can find regional blueberry wines named 'Bleuet', 'Bluet', etc.) 'Bluet' can also refer to various damselfly species belonging to the genera *Coenagrion* and *Enallagma*. 'Blewett,' meanwhile, was a Norman French and Middle English term for blue cloth. After the Norman Conquest, 'Blewitt' (also Bluett, Bloet, Bluiett, etc.) was a name for a person with blue eyes or wearing blue cloth. This is the origin of the modern family surname 'Blewett' and its various other spellings. Regarding the mushroom, 'blewit' derives from *bleuet* (Old French, referring to blue or purple hues).

**Habitat:** *C. nuda* is usually classified as a saprophyte (technically, a white rot fungus) originating from the temperate regions of North America and Eurasia, often found in accumulations of decaying leaf litter in coniferous or deciduous

---

1   The *nudus* part refers to the 'nude' cap, which lacks the *cortina* (web-like veil remnants) of another bluish look-alike mushroom (*Cortinarius violaceus*).
2   Du, J., *et al.* 2018.
3   He, Z. 2023.

forests or also other types of organic matter, debris, or wastes. It has also been introduced to Australia. Its occurrence has been increasing there, often in association with *Eucalyptus* spp. and gorse. (*C. personata* and *C. sordida* are also reported from Australia). It the Northern Hemisphere blewits usually fruit during the later part of the year, i.e. September to December. (In B.C. fruiting tends to peak during late October and November.) Fruiting bodies can be solitary or in groups, or sometimes in fairy rings (as with other related species).

*C. nuda* also has the ability to form mycorrhizal associations with tree roots, which may help to explain its frequent association with forests. This mycorrhizal lifestyle was long suspected[1]—research observations eventually confirmed this, with one study observing mycorrhizal interaction with the roots of Sakhalin spruce (*Picea glehnii*),[2] and another observing the formation of mycorrhizae with Japanese red pine (*Pinus densifolia*).[3] When compared to other mycorrhizal fungi, the flexible lifestyle of *C. nuda* likely provides an opportunistic and/or competitive advantage. The fungus produces enzymes involved in phosphorus and nitrogen mobilization and organic matter decay that are more active in leaf litter compared to those of other mycorrhizal fungi.[4] The microbe might thus live day-to-day in conjunction with its mycorrhizal host tree, but be 'Johnny-on-the-spot' when there is a fresh deposit of fallen leaves.

**Cultivation:** Methods for cultivating blewits have been in development since the late Nineteenth Century.[5] Some North American companies sell kits or inoculum for growing your own blewit patch at home. Blewits can be cultivated on compost in a manner similar to methods used for *Agaricus bisporus*. The fungus will grow on a variety of agar media, for example media incorporating corn meal, malt extract, potato extract and dextrose, or oak extract, often with supplemental nitrogen sources.[6] Sawdust, cereal grains, and bagasse can be used as solid media for producing mycelium and spawn.[7] Fruiting bodies can be produced on compost or straw, where a casing (top surface) layer may or may not be required according to various sources. Cyclic changes in $CO_2$ levels during growth may require ventilation to prevent inhibition of fruiting.[8] Liquid cultures of *C. nuda* (using glucose-yeast extract and oak extract) have also been studied for production of polysaccharides and/or biomass.[9] By crossing different strains, it is possible to use breeding to shorten the time needed to obtain fruiting by around a month.[10] All of this being said, one comparison of blewit cultivation versus that of *Agaricus bisporus* and *Pleurotus ostreatus* found that blewit productivity was not sufficient to establish full-scale commercial operations.[11]

In Asia, methods have also been developed for the cultivation of *C. sordida* on rice straw and other agricultural wastes.[12] One method for a Thai strain was optimized using YM (Yeast-Malt extract) agar for initial growth, sorghum as a solid medium for spawn cultures, and rice straw with a casing layer composed of sandy soil to cultivate the mushrooms.[13]

1   Trappe, J. 1962.
2   Kasuya, M., *et al.* 1996.
3   Chung, H., *et al.* 2002.
4   Colpaert, H. and Van Laere, A. 1996.
5   Costantin, J. and Matruchot, L. 1898.
6   Soon, J. and Bong, K. 2013; Stott, K. 1998.
7   Lee, S., *et al.* 1996; Gaitan-Heernandez, R. and Baez Rodriguez, R. 2008.
8   Stott, K. 1998.
9   Özdal, M. 2018; Gaitan-Heernandez, R. and Baez Rodriguez, R. 2008.
10  Jeon, J., *et al.* 2020.
11  Danai, O., *et al.* 2008.
12  De, J., *et al.* 2022.
13  Thongbai, B., *et al.* 2017.

# Consumption, Nutrition, & etc.

**History:** Older references to wood blewit don't really seem to mention edibility. Bulliard's 1790 description[1] doesn't mention edibility, nor does James Bolton in his 1791 book titled *An history of fungusses growing about Halifax*, wherein he said the mushroom was rare and called it the 'bulbous agaric.'[2] In 1827, a colour lithograph of 'suspect' mushrooms included wood blewit (as *Agaricus nu³ lilas*).[4] Joseph Roques' 1941 French compendium of edible mushrooms doesn't mention it at all.[5] In his 1871 book titled *The Mushrooms of Canada*, D.K. Winder listed the mushroom (as *Agaricus nuda*), but it appears that he did not venture to eat it (indicated by the lack of an asterisk in his list).[6] However, by 1898 cultivation methods were being investigated and in 1975 Jane Grigson reported that the mushroom (she named it *Tricholoma nudum*) was occasionally sold in shops in the British Midland counties, remarking that it was more pleasantly flavoured than a relative she named *Tricholoma saevum*.[7] Currently European supermarkets carry wood blewits in the autumn and winter fruiting season. Common uses include cooking with butter in a simple sauté (perhaps served with meats, cheese, rice, or pasta), or with sauces, omelettes, and stews (especially with leeks and onions).

**Nutrients & Trace Elements:** Nutrients levels reported in *C. nuda* are listed in Table 1 (p. 4) and trace elements are listed in Table 2 (p. 5). *C. nuda* can potentially be a bioindicator of the presence of heavy metals in polluted soils.[8]

**Flavour:** Fresh blewits are said by some to have an aroma resembling frozen orange juice. Others describe a faint aniseed odour. Descriptions of the flavour vary quite a bit. The consensus is that the flavour is pleasant. Some refer to it as indistinct while others disagree and say that it is relatively intense, perhaps even slightly bitter. Other descriptors include sweet, fungal, aromatic, nutty, or mild. This author finds the flavour comparable to *A. bisporus*. Aromatic compounds present in the mushroom include 3-octanone and the two usual suspects (1-octen-3-ol and its flavourful pal, the ketone aptly named 1-octen-3-one—see the detailed discussion in Part 2).[9] The fraction of aromatic monoterpene compounds found in *C. nuda* is dominated by a furanoid compound named cis-linalool oxide (appropriately enough with a sweet, woody fragrance). Other aromatics include trans-linalool oxide (also sweet, woody) and linalool itself (fresh, floral).[10]

**Bioactivity:** Various bioactive properties are reported from laboratory and animal studies of *C. nuda* and are indicated in Table 3 (p. 6).[11] Of particular interest, *C. nuda* produces indole-3-carbaldehyde, a compound shown to completely inhibit zoospore germination in *Phytophthora capsici* (a soil pathogen of peppers and other plants, found in the class of fungi called Oomycetes).[12] It also produces 2-methoxy-5-methyl-6-methoxymethyl-p-benzoquinone, known to be a weak antibiotic and an inhibitor of blood platelet aggregation.[13] It is clear from various sources that **ONE DOES NOT EAT THIS MUSHROOM RAW. IT MUST BE COOKED.** In its raw state the mushroom is considered poisonous by some sources.[14] *C. nuda* protein extracts are insecticidal against fly larvae. This is an excellent property for a mushroom trying not to be consumed by maggots,[15] but laboratory tests show that the extracts are haemolytic (blood-cell destroying).[16] Another source states that the mushroom contains hydrogen cyanide, a toxin neutralized by cooking.[17]

*C. sordida* produces two unique diterpenoid compounds of note: lepistol and lepistal. Laboratory tests have shown that these compounds have activity against leukemia cell lines. In particular, lepistal (an aldehyde) shows greater cytotoxic

---

1   Bulliard, P. 1790.
2   Bolton, J. 1791
3   *Nu* = 'nude' (French)
4   Cornillon, A., 1827.
5   Roques, J. 1841.
6   Winder, D. 1871.
7   Grigson, J. 1975.
8   Kalač, P. and Svoboda, L. 2000; Svoboda, L., *et al.* 2006.
9   Noël-Suberville,, C., *et al.* 1996.
10  Breheret, S., *et al.* 1997.
11  De, J., *et al.* 2022;
12  Chen, J., *et al.* 2012.
13  Lauer, U. and Anke, T. 1991.
14  Before anyone panics, recall that some plant foods also must be cooked in order to be safe.
15  A teleological expression, of course. Science is not actually aware of any mushrooms having the capacity to form ambition. Slime moulds may be another matter, but they are actually social amoebae, not fungi.
16  Wang, M., *et al.* 2002.
17  Gone71 N. 2022.

and antibiotic activity. In liquid culture of mycelia, *C. sordida* also produces nudic acid B, a compound with antibiotic and cytotoxic properties (the compound is not detected in fruiting bodies, however).[1]

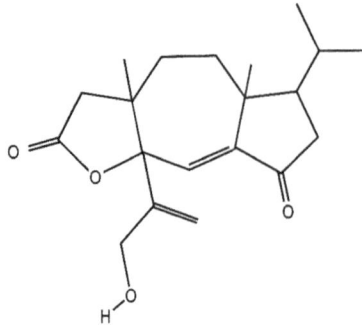 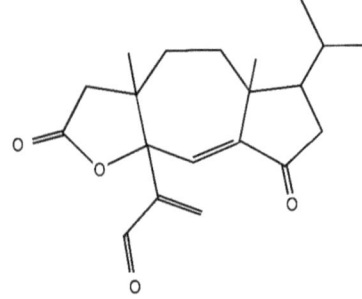

Lepistol                                                                                    Lepistal

**Allergies & cooking:** As noted in the bioactivity section, blewits may cause gastrointestinal distress (sometimes severe) if eaten raw. Even when cooked, some people apparently have difficulty digesting the mushroom or may be allergic to it. If you have never eaten blewits, recall that discretion is the better part of valour (and the admonishment in the introduction to only try a little bit at first and wait an appropriate amount of time (perhaps 24 hours) to see if things are O.K. before consuming more). You read the introduction section of this book before skipping ahead to this section, right?

**Colour:** Most people describe fruiting bodies of *C. nuda* as being light shade of lilac-purple or perhaps bluish-purple, sometimes with tan cap colours. For the author, the colours maybe range somewhere between lilac and periwinkle (periwinkle can be classified as both a shade of blue and a shade of purple). In the context of language and culture, the use of blue vs. purple in some common names probably stems from the surprisingly subjective (and changing) ways that people have perceived and discriminated colours over time (including blue vs. purple).[2] One source asserts that the purple and blue tones of *C. nuda* are due to the presence of phenolic compounds.[3] When fixed with a mordant, pigmentation from blewits can be used to dye fabrics or paper (however, the colour ends up being grass-green—not violet or blue like the fresh fruiting bodies).[4] Interestingly, a white laccase enzyme[5] produced by the fungus can be used de-colourize a number of dyes such as methyl red, coomassie brilliant blue, and reactive brilliant blue.[6] A compound called 6-Hydroxy-2H-pyran-3-carbaldehyde is also produced. It inhibits the enzyme tyrosinase, which means that it can interfere with synthesis of melanin (a pigment found in human skin and also fungal cells).

**Look-alikes:** If you want to hunt for blewits, keep in mind that there are violet and blue-coloured look-alikes. Particular examples include 'chunky' brown-spored members of *Cortinarius* such as *C. violaceus*, *C. traganus* (mildly poisonous), *C. alboviolaceus*, or *C. camphoratus* (potentially poisonous). If one lets the caps rest quietly on a piece of paper for a few hours and the resulting spore deposit is brown or rust-coloured rather than whitish-pink or whitish-pinkish-beige, then the mushroom isn't *C. nuda* and it might be wise to review field identification notes. Hopefully, an expert has already shown you how to identify the field marks that separate blewits from *Cortinarius* species (such as faint traces of a brown or rust-coloured spore deposit on remnants of a web-like *cortina* clinging to the stem beneath the gills of *Cortinarius* spp.). That's not the only field mark to consider, by the way. This book is not a field identification guide—please consult reliable expertise for more complete information about wild-harvesting blewits.

---

1    Thongbai, B., *et al.* 2017.
2    Michaeli, D. 2021.
3    De, J., *et al.* 2022.
4    First Nature. 2022b.
5    Fungal laccase enzymes are responsible for breaking down lignin substrates. This particular enzyme also has antiviral properties.
6    Zhu, M. *et al.* 2016.

# Recipe

**Remember: blewits must absolutely be cooked thoroughly (please read the preceding pages carefully)!**

## Blewit & *Allium* soup

When blewits are mixed in a soup with bold ingredients like leeks, onions, and their *Allium* kin, their mushroom flavour can still hold its own in a very pleasant way. *Allium* is a plant genus that has anywhere from about 260-979 species, and it includes plants cultivated since ancient times. The genus includes garden onions, French red shallots, and scallions (*Allium cepa*), garden leeks (*A. porrum*), cultivated garlic (*A. sativum*), and chives (*A. schoenoprasum*). Onions, in particular, were an important and widely used winter staple for the European settlers of northern North America. Onions were relatively easy to grow, they kept well during the winter, and they could be used either on their own or with other foods to add something savory and nutritious to otherwise bland diets. North American explorers and traders often turned to wild onions (*A. canadense*) for something to make their supplies of pemmican (dried meat solidified in rendered fat) more palatable. In Canada, a stew made with pemmican and wild onions or potatoes was called a 'rubaboo.' The modern day forager's recipe[1] that prompted development of this blewit soup recommended the inclusion of 30-40 wild *A. canadense* cloves (or 16 cultivated garlic bulbs)! That seems rather excessive. Readers will find the recipe on this page produces a kinder, gentler (yet still tasty) soup. If, perchance, you didn't find many blewits, you can supplement them with a few button mushrooms (*Agaricus bisporus*). The flavours aren't 100% identical, but they are very similar in this dish. Again, just make sure you have cooked the blewits thoroughly.

| | |
|---|---|
| 3 tbsp | Vegetable oil |
| 1 | Onion, yellow, chopped |
| 1 | Leek, chopped |
| 5 cloves | Garlic, minced |
| ¼ lb | Blewits (*Collybia nuda*), or *A. bisporus* chopped |
| 2½ cups | Chicken stock |
| 1 cup | Table cream |
| To taste | Salt and pepper |
| 2 tbsp | Parsley leaves, chopped |
| Some | Chives, sliced |

Heat the oil in a soup pot (med. heat) and sauté the onion and leek until limp and translucent (but not browned, perhaps 2-3 min.). Stir in garlic and mushrooms and continue to cooking 5-6 min. until the mushrooms are limp and browned. Add the stock, allow the mix to boil, and reduce heat to simmer ca. 5 min. Stir in the cream, simmer another minute or so, and remove from heat. Add parsley to the mix and puree in a blender. Adjust the salt and pepper seasoning, ladle into bowls, and garnish with chives before serving warm. Serves 3-4.

---

1    Drennan, F. 2010.

# Yellow legs & black trumpets

Artistic depiction of *Craterellus tubaeformis* (by R. Winder). This is not a field guide.

# Biology & Ecology

**Announcing winter:** Finding choice edible mushrooms during winter-time is a true revelation for those used to a hard winter with sub-zero temperatures and snow normally on the ground as early as late autumn. The winter chanterelle (*Craterellus tubaeformis* and its kin) almost seems like it's announcing the arrival of winter with its little yellow trumpet shapes erupting from cold, wet moss and wood before the snow flies. *Tubaeformis* actually refers to that trumpet shape, one that originally convinced mycologists to classify the mushroom as a *Cantharellus*. You'll still find older names like *Cantharellus infundibuliformis* or *Cantharellus tubaeformis* in some literature. However, it seems that taxonomists are always revising things, and DNA tests have resolved the species as belonging to *Craterellus*. DNA studies have also performed their traditional role of splitting up the species and leaving North Americans in the dark. European types are now regarded as the true *C. tubaeformis* and North American types are now regarded as several officially unnamed species. A 2003 work referred to the western North American types as *C. neotubaeformis nom. prov.*[1] If only things could be that simple. Recent work with ITS (internal transcribed spacer) DNA sequences from *C. tubaeformis* distinguished three groups—one from Oregon, B.C., and Québec, one group from Belgium, Sweden, and Denmark, and (the puzzler) one group from Belgium and Québec.[2] So things are currently 'clearly uncertain' and we are reduced to calling the types in western North America '*C. tuabaeformis* group.' More work will be needed to resolve true names. *C. tubaeformis* and its North American friends have lots of different common names. Collectively, the species are often lumped together and called 'winter chanterelles.' *C. tubaeformis* (the most frequently mentioned species) has specific common names that include trumpet chanterelle, funnel chanterelle, yellow legs, yellow foot, winter mushroom, winter chanterelle, *chanterelle grise, chanterelle modeste, chanterelle d'automne* (grey, modest, or autumn chanterelle—France), *chanterelle en tube* (chanterelles in a tube—Québec) *trompetenpfifferling* (trumpet chanterelle—German), *trattkantarell* (funnel chanterelle—Swedish), *suppilovahvero* (funnel wafer—Finnish), *pieprznik trąbkowy* (trumpet pepper—Polish), Zeltkāta gailene (goldenrod chanterelle—Latvian).

Another related group of choice edible *Craterellus* spp. are given common names such as black chanterelle, black trumpet, horn of plenty, or trumpet of the dead. *Craterellus cornucopiodes* is a well-known European black chanterelle that is reported to fruit from August to December. In North America, *C. fallax* (eastern black trumpet) and *C. calicornucopiodes* (California black trumpet) are related species. Recently recognized as a distinct species (in 2015), *C. calicornucopiodes* is rarely encountered in B.C. (but is more common in California).

**Habitat:** Winter chanterelles are mycorrhizal like their bigger chanterelle cousins. They also tend to sprout in large groups or 'troops' from moss and well-decayed 'coarse woody debris' (a forester's technical term for logs, pieces of wood, and larger branches laying on the ground). Descriptions of general habitat for *C. tubaeformis* in northern North America highlight its presence in conifer bogs. In western North America the species primarily forms mycorrhizae with western hemlock (*Tsuga heterophylla*). It will also form associations with Douglas-fir (*Pseudotsuga menziesii*) and Sitka spruce (*Picea sitchensis*), but western hemlock will usually also be present. Abundance of the fruiting bodies increases with the amount of woody debris (on sites where timber was harvested less than a century ago). In B.C. and the Pacific Northwest it usually fruits in late autumn to early winter and is often found in the same habitat as *Hydnum* spp, but not with *Hypholoma* spp.[3] In New England and eastern Canada it fruits from August to November and associate with hardwood or mixed hardwood/conifer forests where dead wood or wood chips and quite often *Sphagnum*-type mosses are present.[4] The mushroom is also associated with diverse forests northern Eurasia (Britain, Ireland, Scandinavia, Russia, or at higher altitudes in areas to the south). In Britain the fungus thrives especially in spruce forests and is abundant in western Wales, northwest England, and Scotland.[5] In Sweden the fungus is found in older forests of Scots pine (*Pinus sylvestris*).[6] The mycorrhizae have also been identified in association with Himalayan poplar (*Populus ciliata*) in temperate forests of Pakistan[7] and the species is also reported to be in India and Thailand.

---

1    *"Nom. prov."* means "provisional name." See Pilz, D. *et al.* 2003.
2    Guo-ying, Z. *et al.* 2017.
3    Trappe, M. 2004.
4    Mushroom-Collecting.com. 2022.
5    O'Reilly, P. 2022.
6    Hagenbo, A., *et al.* 2018.
7    Jabeen, S., *et al.* 2012.

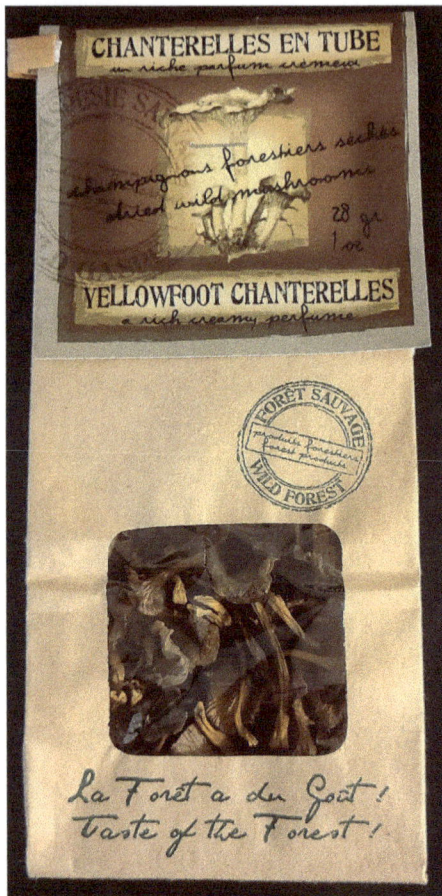

Regarding black trumpets, *C. calicornucopiodes* occurs along the Pacific coast. It is a rare find in B.C., reported to fruit during October and November on the SVIMS checklist.[1] The iNaturalist mobile phone app shows the largest concentration of the species in California, fruiting from November to May (peaking in January). It is reported to favour hardwoods. *C. fallax* occurs east of the Rocky mountains in North America and is mycorrhizal with oaks and perhaps other hardwood tree species. It fruits from May to November (peaking in August) and tends to grow as singletons or in loose clusters (sometimes more gregarious) in mossy areas (particularly *Sphagnum* moss). *C. cornucopiodes* is widely distributed in Europe. It also occurs in eastern North America. It fruits there from June to December (peaking in October) and it favours mossy areas. It associates with beech, oaks, and other hardwood trees.

**Chemical defence?:**

*C. tubaeformis* produces (8E)-10-hydroxydec-8-enoic acid when it is injured (probably deriving it by enzymatic conversion from linoleic acid).[2] There is some speculation that it may help to repel insect attack.[3]

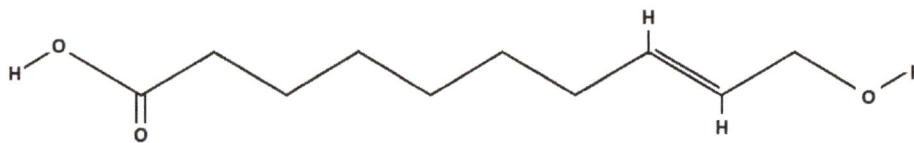

(8E)-10-hydroxydec-8-enoic acid

---

1    Gibson, I. 2020.
2    Pang, Z., *et al.* 1992; Anke, H., *et al.* 1996.
3    Iadanza, N. 2015.

# Consumption & Nutrition

**History of consumption:** The inclusion of winter chanterelles and black trumpets in cuisines follows a trajectory similar to that of previously discussed chanterelles—perhaps there was some traditional or early consumption but sources really don't highlight them until modern times. In Scandinavia, where *C. tubaeformis* is now quite popular, one source describes people of pre-industrial Sweden as "reluctant" or "indifferent" when it came to mushrooms as a source of nutrition. It claims that the peasantry "would eat grass and leaves, sawdust and dirt, almost anything but not mushrooms, preferring rather to starve to death." However, Swedish mycologist Elias Magnus Fries described *C. tubaeformis* as a species in 1821. By 1841, French mycologist Joseph Roques reported on the edibility of winter chanterelles and black trumpets—his statements reflected an emerging awareness of their culinary potential. Roques reported that '*Agaricus infundibuliformis*' had "*une odeur forte, mais agréable*" (a strong but agreeable scent) and that "*On peut l'employer comme aliment*" (it can be used as food). He called black trumpets '*Cantharellus cornucopiodes*' and cited the findings of French mycologist Jean-Jacques Paulet (1740-1826), stating "*Essayée sur les animaux, cette plante ne les a point incommodés, elle a d'ailleurs, dit-il, une légère saveur de truffe*" (Tested on animals, this plant did not bother them; moreover, he [Paulet] says, it has a slight truffle flavour).[1] Back in Sweden, attitudes towards mushrooms evolved. They are now a popular part of modern Swedish cuisine and many households harvest them. Currently, the most popular mushrooms in Sweden are the golden chanterelle (*C. cibarius*), the king bolete (*B. edulis*), and *C. tubaeformis*.[2] Globally, winter chanterelles are often used in a sauté, added to soups, or consumed in other ways that don't overwhelm the flavour (in a béchamel sauce, for example). It is small, hollow-stemmed, and easily dried. That makes the dried form perfect for chefs to stash in a small airtight jar until needed. Winter and black chanterelles are both sold commercially, fresh or in dehydrated form.

**Nutrition & Trace elements:** The literature doesn't really hold much information about the nutritional status of *C. tubaeformis*, probably because it is usually used in relatively modest amounts (more for flavour and perhaps texture rather than being a bulk food). Slightly more information is available for *C. cornucopiodes* (see the summary in Table 1 on p. 4). The mushroom does contain ergosterol (a precursor of vitamin $D_2$). There is an up to nine-fold increase in vitamin $D_2$ content when freeze-dried *C. tubaeformis* is exposed to UV light.[3]

Table 2 (p. 5) lists cesium as a radioactive trace element accumulated by *C. tubaeformis* and *C. cornucopiodes*. Observations of *C. tubaeformis* specimens in Sweden after the Chernobyl accident of 1986 showed that radioactivity reached 522,000 becquerels[4] per kg (with 10,000 Bq/kg considered to be the safe upper limit). Levels in chanterelles subsided to around 100 Bq/kg by 2017.[5] During 2000-2005 (more than a decade after the accident), a Finnish study compared 600 samples of 20 mushroom species in areas of high and low cesium deposition. *C. cornucopiodes* was found to have relatively high levels of radiocesium (ca. 1500 Bq $kg^{-1}$ fresh weight).[6] *Craterellus lutescens* was also reported to highly accumulate radiocesium in Italy after the accident.[7] Given that there were also similar observations for related chanterelles, it would probably be wise to avoid harvest of any of these species where or when there could be any risk of radioactive contamination.

**Aroma, flavour, & pigments:** *C. tubaeformis* has found its way into several studies that compare mushroom flavours and aromas. We can only hit the highlights here. One interesting Finnish study looked at the linkage between non-volatile and volatile compounds and human preferences linked to flavour in button mushrooms (*Agaricus bisporus*), porcini (*Boletus edulis*), curry milkcap (*Lactarius camphoratus*), chanterelles (*Cantharellus cibarius*), and trumpet chanterelle (*C. tubaeformis*). The profile of sugars and organic acids present in *C. tubaeformis* was similar to that of *C. cibarius*. *C. tubaeformis* did have relatively higher amounts of mannitol and much lower amounts of trehalose compared it its larger cousin. The results were interesting when preferences were examined. There was more variability linked to different groups of people tasting these mushrooms than there was linked to different species.[8] In another study, *C. tubaeformis* and its larger *C. cibarius* cousin had relatively high levels of bitter-tasting amino acids (e.g. L-arginine, L-histidine, and L-tyrosine).[9]

---

1    Roques, J. 1841.
2    Svanberg, I. and Lindh, H. 2019.
3    This was achieved with incubation of up to 2 hours at 250 nm and 20 cm distance from the UV light source. Please see Jiang, Q., *et al.*, 2020.
4    A becquerel is a measure of radioactivity. One becquerel is one radioactive decay per second.
5    Masood, H. 2022.
6    Kostiainen, E. 2007.
7    Battison, G., *et al.* 1989.
8    A good point to remember when someone says that they don't like a particular mushroom species. Preference is sometimes more about the consumer than that which is consumed. See Aisala, H., et al. 2020.
9    Manninen, H., *et al.* 2018.

Over a dozen volatile compounds have been detected in *C. tubaeformis,* with the highlights including 1-octen-3-ol and 1-octen-3-one (mushroom aroma)[1], 2-octen-1-ol (mushroom-like aroma), 6-methyl-5-hepten-2-one (green, mushroom-like aroma), limonene (lemon-like aroma), nonanal (fruity, sweet, pine-like aroma), linalool (citrus-like aroma), 2-phenylethanal (lilac/hyacinth-like aroma), benzaldehyde (burnt sugar, almond aroma), and benzyl alcohol (sweet-spicy aroma).[2] A study of flavour compounds formed during heating (cooking) and browning of the mushroom extract (known as the 'Malliard reaction') correlated 3-phenylfuran and 2-octylfuran with caramel flavours, 1-octen-3-ol, (E)-2-octen-1-ol and geranyl acetone with mushroom-like flavours, and 2-thiophene-carboxaldehyde, along with 2,5-thiophenedicarboxaldehyde and 3-methylbutanal with improving meat-like flavours. The optimal temperature for a flavourful browning reaction in *C. tubaeformis* was found to be 125°C.[3]

The flavour of black trumpets is regarded as choice, putting them in the same category as king boletes, morels, and truffles. Without regard to particular species, people describe a complex umami flavour and rich, fragrant earthiness and/or smokiness. The flavour may be described as having notes of, for example, truffles, chocolate, nuts, and/or sweet fruitiness (of apricots or stone fruit). Some also find that the mushrooms have a slightly bitter aftertaste. This author would describe the flavour as something that catches you a bit off-guard if you've been trying other edible species. It's complex and pleasing, not in a timid way but in an elegant, super-friendly way that says 'Now *this* is a mushroom!' Other mushrooms should probably be compared to black trumpets, and not *vice versa.* My own tasting experience matches the flavours mentioned above, except that the aftertaste seems to be be pleasantly sharp and/or tart rather than simply bitter. All of this pleasing complexity can probably be put down to a group of compounds that are the usual suspects for umami flavours: glutamates and 5' nucleotides.[4]

Winter chanterelles in a bowl. This is not a field guide.

The yellow pigmentation of *C. tubaeformis* may derive from carotenoid pigments.[5] The mushroom can be used to dye textiles and paper. It produces a yellow colour when ammonia is used as a mordant.[6] *Craterellus* spp. also produce black pigments (notably in black trumpets),[7] but work identifying the compounds could not be located for this summary.

**Bioactivity:** *C. tubaeformis* have been shown to be anti-inflammatory in laboratory tests of cell cultures (Table 3, p. 6).[8] A Finnish study of estrogen-like contents in foods has listed *C. tubaeformis* as containing two lignans: matairesinol (7.3 μg/100g fresh weight) and secoisolariciresinol (3.9 μg/100g fresh weight). To put this in context, the average daily consumption of phyto-estrogen compounds in Finland was estimated to be 500 μg.[9]

*C. cornucopiodes* produces compounds reported to have a wide range of activities, including immunomodulatory, antibiotic and anti-viral, anti-inflammatory, anti-diabetic, anti-cholesterol, anti-allergic, cardiovascular, anti-cytotoxic (organ protective) and antioxidant activity (Table 3, p. 6).[10] It is reported to have antibiotic activity linked to compounds

---

1    See also Part 2 for more on the importance of 1-octen-3-ol in mushroom flavours.
2    Fons, F., *et al.,* 2003; Chen, X., *et al.* 2018a.
3    Chen, X., *et al.* 2018b.
4    Beluhan, S. and Ranogajec, A. 2011.
5    Faisson, J. and Arpin, N. 1967
6    Mushroom-Collecting.com. 2022.
7    Arpin, N. and Fiasson, J. 1971.
8    O'Callaghan, Y., *et al.* 2015.
9    Valsta, L., *et al.* 2003.
10   Boruah, A. 2023; Ozuna-Valencia, K. *et al.* 2023.

that include quercetin, ferulic acid, gallic acid, and *p*-coumaric acid.[1] Anti-oxidant and anti-diabetic properties are linked to a compound called myricetin.[2] Interestingly, the antioxidant and antimicrobial properties of this mushroom were found to enhance shelf life when it was incorporated into sausages (although colour and flavour were negatively impacted).[3] *C. cornucopiodes* also produces several sesquiterpenoid compounds (craterellins A-C and gymnomitr-3-en-10β,15-diol), among which craterellin C is reported to be cytotoxic against one tumour cell line.[4]

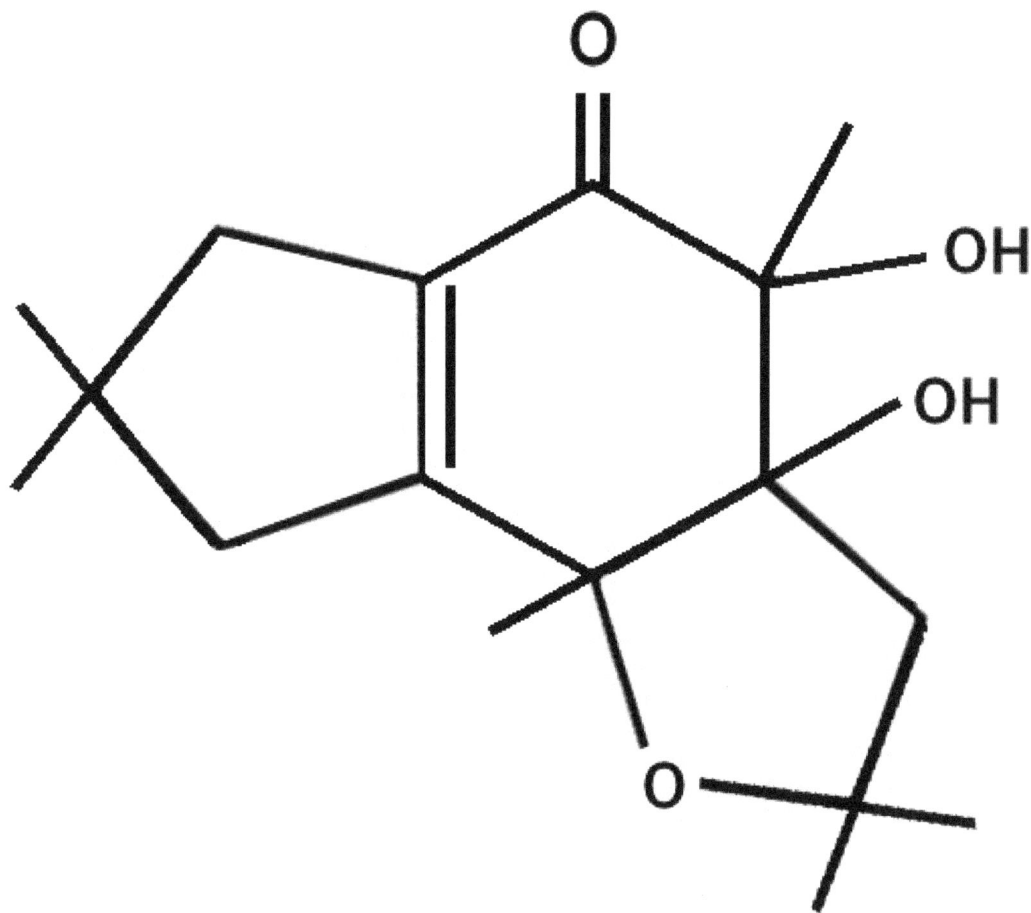

Craterellin C

**Look-alikes:** The main look-alikes for *C. tubaeformis* are other winter chanterelles and their *Craterellus* kin (but this does not excuse the reader from following the admonition sprinkled throughout this book—consult appropriate expertise and learn to identify before attempting a wild harvest). Having various distributions, other winter chanterelles include *C. lutescens* (a.k.a. yellowfoot, widely distributed), *C. ignicolor* (a.k.a. yellowfoot or flame-coloured chanterelle, found in eastern North America), *C. odoratus* (fragrant chanterelle, found in the southeastern U.S. and Mexico), and *C. lateritius* (smooth chanterelle, found in Asia, Africa, and North America).

1   Kosanić, M. *et al.*
2   Ozuna-Valencia, K. *et al.* 2023.
3   Novakovic, S. 2021.
4   Thu, Z., *et al.* 2020. Guo, H. *et al.* 2017.

# Recipes

## Trattkantarell soup

*Trattkantarell* is the Swedish name for funnel mushrooms[1] (a.k.a. *Craterellus tubaeformis* group., yellow legs, winter chanterelles). This is my North American take on a Scandinavian recipe[2] for funnel chanterelle soup. The fairly standard original soup ingredients are changed here to include summer savory, which always goes nicely with a chicken stock. Also, Icelandic *skyr* is used here to enhance the creamy flavours. Skyr resembles Greek yogourt, but it's a rich curdled sour milk cheese derived simply from native cultures (where Greek yogourt is enriched by straining whey liquids from regular yogourt). Using just a little adds a nice, slightly sharp edge to the flavours. You don't need to go all the way to Reykjavik to get it—skyr is available in B.C. grocery stores. Inspired by a similar recipe for 'yellowlegs soup' appearing a recent issue of *Fungi* magazine,[3] North American wild rice is also included. Enjoy your lunch!

| | |
|---|---|
| 2 cups | Warm water |
| 1 cup | Funnel chanterelles (*C. tubaeformis* grp.), dried |
| 4½ cups | Chicken stock |
| 3 tbsp | Butter |
| 1 | Onion, chopped |
| 2 cloves | Garlic, minced |
| 2 tsp | Thyme |
| 2 tsp | Summer savory |
| ½ cup | Rice, long grain, brown |
| ¼ cup | Wild rice |
| 3 tbsp | Skyr |
| ¼ cup | Cream |
| To taste | Salt & pepper |

Remove tougher stems from the dried chanterelles and rehydrate tops in warm water 30 min. Drain (while reserving the water to make the chicken stock). Melt butter in a large stew pot (med. heat) and sauté mushrooms until limp and browned. Add onion, garlic, thyme, and summer savory, and continue cooking and occasionally stirring ca. 2 min. until onion is soft and transparent and garlic aromatic. Add the chicken stock, bring to a boil, and reduce heat to a simmer. Add the rice, cover, and simmer ca. 50 min. (until the rice is tender)[4]. Mix the skyr and cream well and stir that in, season with salt and pepper, and remove from heat. Serve warm in bowls and garnish with a dollop of skyr. Serves 4-5.

---

1    In case you favour a different Northern European country, it's *traktkantarell* in Norwegian, *tragtkantarel* in Danish, *trektarkantarella* in Icelandic, and *suppilovahvero* in Finnish.
2    Helena. 2023.
3    Towns, K. 2024.
4    Note that the cooking for long-grain brown rice more closely matches that of wild rice. White rice would become mushy if cooked this long.

## Black trumpet tart

This tart is adapted from a published recipe for 'woodland mushrooms'[1] (walnuts are omitted). Fresh black trumpets are choice but hard to find—these tarts were made with rehydrated mushrooms that were nevertheless tasty.

| | |
|---|---|
| 24 | Tart shells, frozen |
| 4 | Eggs |
| 2 cups | Black trumpets (*Craterellus fallax*) |
| 1-2 tbsp | Butter |
| 1½ cups | Whipping cream |
| ½ cup | Goat cheese, crumbled |
| To taste | Salt & pepper |
| 1 tbsp | Thyme, minced |
| 1 tbsp | Italian parsley, minced |
| 1 tbsp | Basil, minced |

Remove tart shells from freezer and thaw (5-10 min.) Preheat oven to 375°F. Beat an egg and brush it inside the shells. On a baking tray, put shells in the oven and bake about 10 min While the shells are baking, melt butter in a skillet (med. heat). Chop and sauté mushrooms until limp and browned (6-8 min.). Remove shells from oven and set aside to cool. Lower oven temperature to 325°F. Whisk together remaining eggs and cream in a bowl until they thicken (3-5 min). Divide the chanterelles between the tarts, followed by a spoonful of cheese, a pinch of the thyme, Italian parsley, and basil, and a tiny dash of salt and pepper. Pour the egg mix to cover contents in each shell. Return tarts to the oven and bake ca. 25-35 min. until centres are firm.

## Winter chanterelle tart

This winter treat combines the method at left with another published recipe[2] to produce an elegant tart to enjoy after laying your hands on a nice batch of winter chanterelles (or celebrating the arrival of your dried ones in the mail).

| | |
|---|---|
| 24 | Tart shells, frozen |
| 4 | Eggs |
| 2 cups | Yellow legs (*C. tubaeformis* group) |
| 1-2 tbsp | Butter |
| 1-2 | Shallots, chopped |
| To taste | Salt &pepper |
| 1½ cups | Whipping cream |
| ½ cup | Emmenthal cheese, shredded |
| 1 tbsp | Thyme, minced |
| 1 tbsp | Summer savory |

Thaw tart shells (5-10 min.) Preheat oven to 375°F. Beat an egg and brush it inside the shells. On a baking tray, put shells in oven and bake about 10 min. While the shells are baking, melt butter in a skillet (med. heat). Chop and sauté mushrooms until limp and browned (6-8 min., until browned, not burnt). Season with salt & pepper, add shallots, and cook until soft and translucent. Remove from heat and remove shells from oven to cool. Whisk together remaining eggs and cream in a bowl until thickened (3-5 min). Divide chanterelles and shallots between the tarts, then add a spoonful of cheese, a pinch of the thyme and summer savory. Pour the egg mix to cover contents in each shell. Return tarts to oven, bake ca. 25-35 min. until centres are firm.

---

1    McEwan, M. 2021.
2    Reder, H. 2020

# Yellow leg omelette

Wake up sleepy-head! Winter is here! And so is your stash of dried yellow legs. It's time to brew a cup of coffee and make a country-style omelette. This recipe is tweaked from one on page 182 in D. Fischer and A. Bessette's 1992 book titled *Edible Wild Mushrooms of North America*.[1] The identification and taxonomic section has been overtaken by more recent information, but the recipe section is still excellent. Anyway, it's time to get those eggs out of the fridge (or out of the chicken coop, if you live at Chez Winder), grab your skillet, and polish up your omelette-making skills. Now you're just showing off!

| | |
|---|---|
| 2 cups | Warm water |
| 1 cup | Yellow legs, dried (*C. tubaeformis* gp.) |
| ¾ cup | Bell pepper, diced |
| 2 | Shallots, diced |
| 4 tbsp | Butter |
| ¼ tsp | Mace |
| ½ tsp | Salt |
| ¼ tsp | Pepper |
| ⅛ tsp | Cayenne[2] |
| 4 | Eggs |
| ¼ cup | Cream |
| ½ cup | Cheddar cheese, mild, shredded |
| 2 sprigs | Parsley, chopped |

Snap the tough stems from the dried mushrooms and rehydrate them in warm water 30 min. When the mushrooms are nearly ready, melt 2 tbsp butter in a skillet (medium heat) and sauté the green pepper and shallots until the shallots are soft and partly translucent. Drain the mushrooms, add them to the skillet along with the mace, salt, pepper, and Cayenne, and continue to cook (while stirring occasionally until the mushrooms are softened and browned, perhaps 5-6 min. Remove them from the skillet, set aside in a bowl, and cover to keep warm. Melt remaining butter in the skillet with reduced heat (medium or lower). Whisk together eggs and cream and pour them into the skillet. As the edges begin to set, use a spatula to gently lift them, allowing liquid egg to flow in. Keep doing this as more egg sets. When it is all nearly set, sprinkle the cheese on top and layer on the sautéed mushroom mix and the parsley. When the eggs are fully set (perhaps another minute) fold the omelette over. To plate, grasp the skillet so that the handle is pointing away from you and the pan is oriented *toward* you. Hold the plate close underneath as you rapidly invert the pan to drop the omelette. Serve warm to 1-2 people.

---

1    Fischer, D. and Bessette, A. 1992.
2    You can omit this if you find spicy seasonings to be too much, but consider offering other diners a bottle of hot sauce on the side in case they prefer it.

# Scottish-style beef & trumpets

*C. tubaeformis* is primarily a summer mushroom in Scotland (where trumpet chanterelle is one of its common names). But members of the *C. tubaeformis* group are winter mushrooms in B.C., so where to put this recipe? Purists will just have to roll their eyes, sigh deeply, and accept that it is listed here in the winter section. Purists will also wonder what makes this recipe Scottish. It's actually adapted from a dish for beef and 'chanterelles' appearing in a compilation of traditional Scottish cookery.[1] The highlight is the way that this thrifty (and perhaps less tender) cut of beef is prepared —seared in strips with some butter added in towards the end. Yours truly has adjusted the published recipe by specifying a mix of *C. tubaeformis* and *Agaricus bisporus* as ingredients. Things are also tweaked a bit by using plum wine rather than dry white wine (although there's probably nothing wrong with using regular white wine). Regardless of which wine you choose, the trumpet chanterelles will take up and concentrate the flavour. That means that the mushroom flavour won't be the focus, but instead contribute some wild and savory notes to this very flavourful dish. Many recipes work well by mixing *A. bisporus* with wild-harvested mushrooms. That's something worth considering for reasons beyond bolstering the flavour and adding some complexity. Buying wild mushrooms can be expensive and using mixed mushrooms can help to stretch your budget. It's also possible that increasing demand could one day outstrip the supply of wild mushrooms and ruin a good thing. Employing a little finesse and only using what you need to elevate a dish is not only good for your bank account—it's a great skill to learn that might also be good for the planet. As an added bonus, it also annoys the heck out of those purists (and that's always great fun!).

| | |
|---|---|
| 4 tbsp | Olive oil, extra-virgin |
| 3 6-oz. | Round steaks,[2] cut into thin strips[3] |
| 2 | Shallots, chopped fine |
| 2 cloves | Garlic, crushed, minced |
| 2 cups | *Agaricus bisporus*, sliced |
| 0-1 cup | Trumpet chanterelles (*C. tubaeformis* grp.) |
| 4 tbsp | Plum wine[4] |
| 4 tbsp | Cream |
| 1-2 tsp | Parsley flakes |
| 2 tbsp | Butter, melted |
| To taste | Salt & pepper |

Add half of the olive oil to a skillet and heat it (medium heat). Sauté the beef a few strips at a time. Brown the strips on all sides (pour off any excess liquid that appears), and set aside (keep warm). Add the rest of the oil to the skillet and sauté the mushrooms until they are brown, soft, and just beginning to caramelize (perhaps 4-6 min.). Add the shallots and continue to cook until the pieces are soft and translucent (1-2 min.). Add and deglaze with the wine, scraping any fond from the bottom of the pan. The wine will probably sizzle right away, so reduce the heat to a low simmer and stir in the cream and parsley flakes. Cook 1-2 more min. and when liquid has reduced to the desired thickness, add back the meat. When the meat is nicely warmed up again and coated, remove from heat, swirl in the melted butter without stirring very much, plate and serve warm. Serves 3-4.

---

1    Wilson, C. and Trotter, C. 2005. The 'chanterelles' depicted Wilson and Totter's recipe actually appeared to be *C. tubaeformis*.
2    You might find that more tender cuts of beef (or venison) will work even better. But round steaks are still very nice.
3    Pat the meat dry with paper before slicing. Round steaks can be a bit challenging to slice. If you have an ulu, use it to push down and cut the meat as deeply as possible, and then draw a chef's knife through the cuts to finish making a clean slice.
4    Suggested recipe in Part 1.

# Venison steak & black trumpets

In 2002, I was stuck in a small hotel in Oslo, Norway during an August heatwave. On my own and needing get out into the cooler nighttime air, I roamed downtown looking for a good place to have dinner. I found a small restaurant offering reindeer steak with chanterelles and lingonberry sauce. That delicious meal serves as the inspiration for the dish described on this page. Canadian caribou are related to reindeer, but they are not farmed (and some subspecies are endangered). Black-tailed deer, on the other hand, are clearly not endangered (just ask any gardener in B.C.). Venison is available at some Canadian markets, so that will be featured here. For the mushroom sauce, I thought about using winter chanterelles. But while shopping for venison at a local market, behold! One last, lonely packet of dried black trumpet mushrooms was right there on a shelf in the produce section! Black trumpets are very tasty in sauces used with meat.[1] It had to be a sign from the heavens. I already had a jar of lingonberry sauce, so I was good to go. Lingonberries (*Vaccinium vitis-idaea*) are popular in Central and Eastern Europe, and a staple item in Northern Europe (especially Sweden). Cranberries a relatively tart alternative to the more refined lingonberry flavour (in my humble opinion).

## Venison & mushrooms

| | |
|---|---|
| 3 | Venison steaks[2] |
| 2 tbsp | Olive Oil |
| 1-2 | Shallots (chopped) |
| ½ cup | Black trumpets (*C. fallax*), dried |
| ¼ cup | Wine (syrah) |
| 1½ cups | Beef stock |
| 2 tsp | Worcestershire sauce |
| 3 tbsp | Skyr (see p. 20) |
| ¼ cup | Milk, 3½% |
| To taste | Salt & pepper |
| Some | Lingonberry sauce |

## Creamed carrots[3]

| | |
|---|---|
| Some | Water |
| 1 lb | Carrots, Julienne-cut |
| 2 tbsp | Butter |
| ½ | Onion, diced |
| 2½ tbsp | Flour |
| To taste | Salt & pepper |
| 1 tbsp | Sugar |
| 1¼ cups | Milk, 3½% |

---

1    Fisher, T. 2024; Mallet, E. 2024.
2    The mushroom sauce also works very well with roast beef.
3    Bobeck, M. 2024.

Rehydrate the dried trumpets in warm beef stock ca. 30 min. Remove and drain (conserve the stock), setting them aside in a bowl. Put the carrots in a pot, add enough water to cover them, and bring the water to a boil. Reduce heat to a simmer and cook until tender (ca. 20 min.). Meanwhile, sear the steaks in a skillet and cook until they reach your preferred level of doneness. Thick-cut venison may take about 9 minutes on each side, at lower heat than you use for beef (to avoid burning). Set steaks aside on a plate, cover, and keep warm. Add olive oil to the skillet, sauté the shallots until they are limp and translucent (2-3 min.). Add mushrooms and cook ca. 5 min. Deglaze with the wine, scraping any fond from the bottom of the skillet. Add the stock and Worcestershire sauce, and reduce heat to a simmer. Cook until the liquids are half-reduced (ca. 15 min.). Meanwhile, melt the butter in a saucepan (med.-high heat) and sauté the diced onions 2-3 min. until tender. Stir in the flour to make a roux, season with salt, pepper, and sugar. Slowly add the milk while stirring, until the béchamel sauce is thick and starts to boil. Drain the carrots and mix them with the sauce. Back at the skillet, mix the skyr and milk very well, stir it into the skillet, and adjust seasoning (salt and pepper). The skyr may curdle some. Simmer 5-6 min. until sauce thickens. Plate warm venison with mashed potatoes and mushroom sauce on top. Place creamed carrots alongside, along with a few spoonfuls of lingonberry sauce where it looks best. Serves 3.

# Hedgehog mushroom

*Hydnum sp.*[1]

---

1    Engraved illustration of *Hydnum repandum* – "Spine-bearing mushroom" from Winder, D. 1871. This is not a field guide.

# Biology & ecology

**Hedgehogs:** Hedgehogs (members of the mammalian family Erinaceidae) are not porcupines (rodents in the Erethizontidae), and neither are hedgehog mushrooms (comprising some *Hydnum* species belonging to the Hydnaceae, a family in the same order as chanterelles, the Cantharellales). *Hydnum* species have a unique spore-bearing surface composed of blunt projections variously described as teeth, spines, spikes, etc. Much like the spines on a hedgehog, the teeth on a hedgehog mushroom are rather short (and they don't detach). That resemblance leads to common name for some of these edible types. *Hydnum* species come in various shades and sizes, and some are quite tough and/or bitter and therefore inedible. The edible ones often surprise mushroom hunters. They can resemble a chanterelle cap when viewed from above, but when picked and turned over—teeth! In B.C., hedgehogs tend to appear in late fall and early winter as colder temperatures hit. If hard frosts come or snow falls they may take a pause. But it's not uncommon to once again find them in January, growing even at the edges of melting snow banks.

The scientific names of hedgehog mushrooms are still being sorted out. We definitely know there are some species and we definitely know that some of them have some names. Putting the right names on the right species and new names on the ones without names is another question. Starting first with the genus, you might run across the names *Dentinum repandum* and *Dentinum umbilicatum* for hedgehog mushrooms in some literature. Without going too far into the weeds that are the rules of the International Code of Botanical Nomenclature, the *Dentinum* name comes from the descriptions of English botanist and mycologist S.F. Gray in 1821. However, Swedish botanist Elias Magnus Fries published his *Systema mycologicum* the same year. In this work, Fries is assumed to have 'sanctioned' the names given by another famous Swedish taxonomist, Carolus Linnaeus, in Linnaeus's 1753 *Species plantarum*. Linneaus had adopted a 1717-1719 description of a monotypic (one species) genus called '*Erinaceus*' by German botanist Johann Jakob Dillenius. But Linnaeus renamed it to *Hydna* and then *Hydnum*, and added some species to the genus. Unfortunately for Gray, Fries published his work a bit earlier in real time. That means that the description of Dillenius as reorganized by Linnaeus as sanctioned by Fries has priority. Phew!

If scientific quarrels over the name of the genus gave you headache, you might want to reach for an analgesic remedy because we now have to consider the names of species. Until recently field guides generally mentioned *Hydnum repandum* and *H. umbilicatum* as edible species present in British Columbia and western North America. Thanks to advancements in DNA-related research we can now clarify this situation as being more unclear. *H. repandum* (a.k.a. sweet tooth, wood hedgehog, or *pied-de-mouton*—French) is not (or probably not) in North America in the opinion of some experts.[1] One study of *Hydnum* species used, in part, ITS1 and ITS4 segments of DNA to look at *Hydnum* species around the world. It assigned *H. repandum* to a '*Hydnum*' subsection of the genus *Hydnum*, and found there were no members of the subsection in North America, other than the similar species *H. neorepandum*.[2] *H. washingtonianum* (the American hedgehog, known to occur in B.C.) and *H. olympicum* (in *H.* subsect. *olympica*) are other species mentioned by various sources. On the other hand, an earlier analysis of *Hydnum* species using ITS1, 5.8s, and ITS2 DNA sequences did detect the presence of *H. repandum* DNA associated with the root tips of Bishop pine (*Pinus muricata*) in California (despite not detecting it in actual North American mushrooms). Meanwhile there are clear indications of other cryptic species in the region, as yet undescribed and unnamed but showing up in the DNA work.[3] Given the known issues with ITS precision and in lieu of further clarity, this work will simply refer to things that look like *H. repandum* in North America as *Hydnum neorepandum* and allies. Hedgehog mushrooms in North America with a dented cap used to be called *H. umbilicatum* (a.k.a. depressed hedgehog or sweet tooth), but DNA work has once again risen to the challenge, found differences between the original European species and species in western North America, and found signs that there are actually more than one unnamed and undescribed species in the region being lumped under the same name. *Hydnum oregonense* (in *H.* subsect. *tenuiformia*) is one of the species falling with this group.[4] Until the other species are named, this work will refer to the western North American type as *H. oregonense* and allies.

---

1    Kuo, M. 2020; MacKinnon, A. and Luther, K. 2021.
2    Niskanen, T. *et al.* 2018.
3    Feng, B., *et al.* 2016.
4    Wood, M. and Stevens, M. 2021.

**Habitat:** Hedgehog mushrooms are mycorrhizal. In B.C. their habitat often coincides with that of chanterelles, fruiting singly or in groups. You might find them as early as late summer but mushrooms usually begin to proliferate in mid-autumn to mid-winter. In western North American *H. neorepandum* and allies can be found in association with conifers like Douglas-fir (*Pseudotsuga menziesii*) or Bishop pine, while *H. oregonense* and allies can associate with those trees as well as western hemlock and various hardwoods, often near stumps and decaying logs. In Europe, *H. repandum* can associate with hardwoods and conifers. *H. umbilicatum*, meanwhile, is found in boggy conifer (pine) forests in Europe and eastern North America.

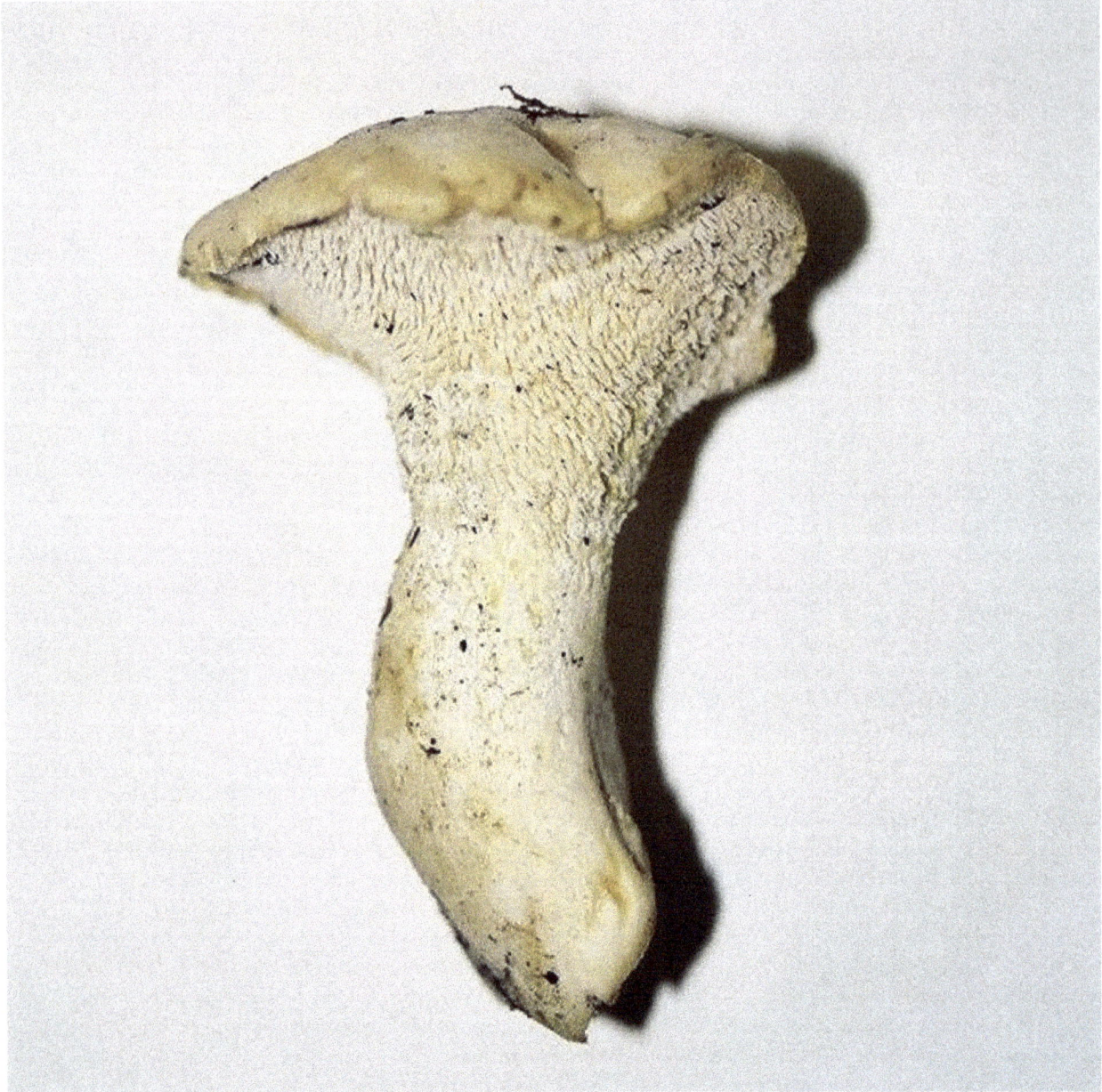

An exceptionally pale hedgehog mushroom (*Hydnum* sp.). This is not a field identification guide.

# Consumption & nutrition

**History:** Hedgehog mushrooms were probably eaten traditionally in some areas of Europe and perhaps elsewhere. With the advent of better botanical descriptions and the Industrial age, wider interest in the mushroom began to pick up steam.[1] In his 1832 work titled *Histoire des champignons comestible et vénéneux*,[2] Joseph Roques described the use of the mushroom under the names *eurchon* or urchin, *erinace*, *rignoche*, and *pied de mouton* blanc (white sheep's foot) in some French departments, and under the name *penchenille* around Toulouse. He also mentioned its use in Austria and the Soigne forest in Belgium, and in Italy under the names *stecchenno*, or *deniino dorato*. Roques wrote:

> «*Mais qu'on se rassure. Les épreuves que j'ai souvent faites sur moi-même avec ce champignon, son usage généralement répandu en France, en Italie, en Allemagne, etc., ne laissent aucun doute sur ses bonnes qualités.*»

> "But let's be reassured. The tests that I have often made on myself with this fungus, its use generally widespread in France, Italy, Germany, etc., leave no doubt as to its good qualities."

In his discussion of *H. repandum*, Roques also provided some culinary advice:

> "As well as some other species of a firm texture, this one needs a prolonged cooking. We cut it into pieces that we pass in boiling water, and which is then cooked with lard, pepper, salt, parsley and broth. This is how I have these mushrooms prepared for my use. We can moreover prepare them with butter, olive oil, poultry fat, a hint of garlic, and a little verjuice (vinegar made from unripe grapes) or lemon juice."[3]

By 1871, D. K. Winder was quoting Roques' assessment in his brief work titled *The mushrooms of Canada*[4]—clearly interest in consuming hedgehogs had been transplanted to North America even as it was growing in Europe.

**Nutrition & trace elements:** Standing as an example (since it has received the most attention of hedgehog mushroom species), the nutrient profile of *Hydnum repandum* is shown in Table 1 (p. 4). *H. repandum* was determined to be a heavy accumulator of radioactive cesium (Table 2, p. 5)[5] after the Chernobyl accident of 1986. In accord with the guidance already given for chanterelles in this work, the better part of wisdom would be to avoid consumption of the mushroom in areas or periods where radioactive contamination could be a concern.

**Flavour & aroma:** The flavour of hedgehog mushrooms is variously described as mild, sweet, nutty, meaty, savory, like chanterelles but without the apricot-like aspect. There may be a hint of bitterness or a peppery aspect. They may acquire a bit of a mineral flavour as they age, and the dried form is said to be less preferable than fresh (so freezing is reccomended rather than drying for preservation). That being said, it is possible to purchase them in dehydrated form. Predominant volatile compounds reported from *H. repandum* include 2-phenylhex-2-enal, that reliable pillar of mushroom flavour 1-octen-3-ol, nonanal, hexanal, (E,E)-deca-2,4-dienal, (Z, E)-deca-2,4-dienal, benzoic acid, decanal, (E)-oct-2-enol, octan-3-one, and others.[6]

---

1    Obligatory pun.
2    Translation: History of edible and poisonous mushrooms.
3    Roques, J. 1832.
4    Winder, D. 1871.
5    Kalač P. 2001.
6    Fons, F., *et al.* 2003.

**Bioactivity:** Unsurprisingly, researchers seeking leads for potential pharmaceutical products have detected a variety of bioactive properties in laboratory analysis of hedgehog mushrooms. Most studies reference *H. repandum* (Table 3, p. 6),[1] although no attempt will be made here to reassign the results to current taxa as we understand them (in other words, the studies name *H. repandum* but they might or might not apply to other hedgehog species—your mileage may vary). Studies of *H. repandum* highlight antioxidant activity linked to the relatively high content of phenolic acids (e.g. ferulic, *p*-coumaric, chlorogenic, gallic and syringic acids) and flavonoids (quercetin, catechin and rutin) found in the mushroom. The variety of compounds detected in *H. repandum* includes sarcodonin A, an anti-inflammatory diterpenoid also found in other *Hydnum* spp.[2] Interestingly, one study found that while extracts of the mushroom had cytotoxic activity against cancer cell cultures, but also had protective action against mitomycin, a cytotoxic anti-cancer drug. Another study has implicated piceatannol as a cytotoxin in Cantharellales (including *H. repandum*).[3] Given the detection of cytotoxic and mutagenic effects in the mushroom, consumers would be well advised not to consume the mushroom in raw form.[4]

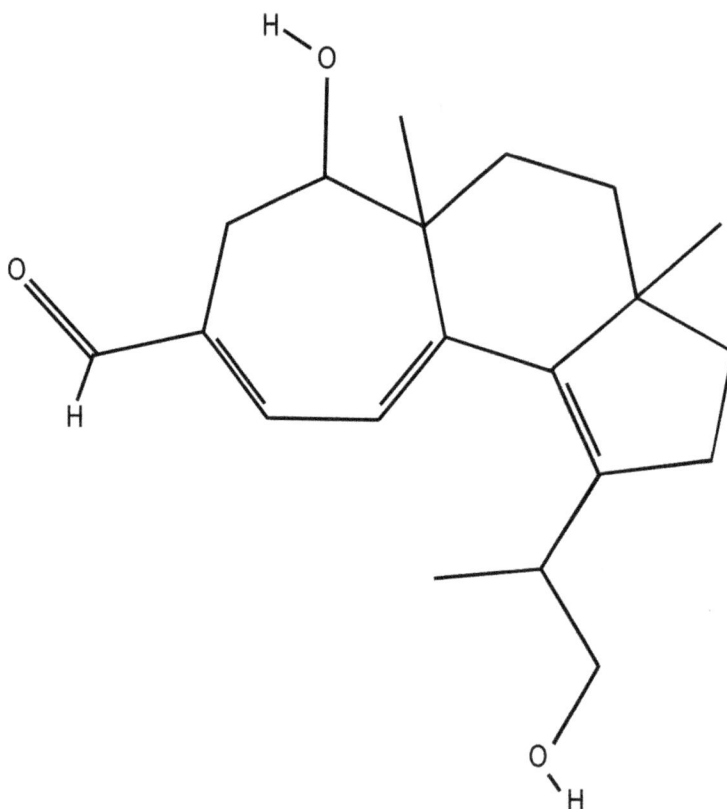

Sarcodonin A. Note presence of interlocking 5-, 6-, and 7-carbon rings

**Lookalikes:** While it is difficult to confuse toothed mushrooms with gilled or pored mushrooms, there will always be someone that tries. There are also *Hydnum* species and other fungi with teeth (often tough or bitter or insipid) that aren't edible hedgehog mushrooms. So while confusion of hedgehog mushrooms with look-alikes is relatively infrequent, let's go ahead and remind ourselves not to undertake a wild harvest for the first time without consulting appropriate expertise. There are a few *Hydnum* species have 'hedgehog' somewhere in their common name. In North America there is *H. albidum* (white hedgehog) and *H. albomagnum* (giant hedgehog). In Europe there is *H. rufescens* (terracotta hedgehog).

---

1    Vasdekis, E., *et al.* 2018; Tubić, J. *et al.* 2019; Deo, G., *et al.* 2019.
2    Liu, J. 2007.
3    Vasdekis, E., *et al.* 2018.
4    Tubić, J. *et al.* 2019.

# Recipes

## Finnish baked hedgehog mushrooms

On July 1, 2009, I attended a banquet dinner during a week-long conference in Helsinki. There were hundreds of us dining at the Old Student House (Vanha Ylioppilastalo), an elegant Neo-Rennaissance building with a stage at one end of its large, central, gilded hall. Meeting organizers had arranged for some attendees to rehearse some songs, and after practising that evening they formed a small choir that would sing for us. I sat at at table with other Canadians and we relaxed as we enjoyed a short serenade before the meal started. We enjoyed it, that is, until the conference organizers approached our table. Someone had informed them that this was Canada Day. Wouldn't it be wonderful if we would go up on stage and perform *O Canada* for everyone? No, really. Impromptu. For a large international audience. Please! After much resistance and gentle arm-twisting, they dragged us onto the stage and we had to sing. Well then! After a tenuous start, my high school choir experience took over. I boomed and belted out the tune as loudly as I could into the cavernous space before us and everyone on stage followed. We all managed to finish valiantly and in unison (with the Americans in the front row rolling their eyes in derision). The True North strong and free... Asked afterwards about our performance, the Finns thought our singing was 'so-so,' but they were truly impressed that we knew the lyrics (they weren't sure they could so confidently perform the words for their own anthem). We were the darlings of the evening. Anyway, this Finnish treat[1] will also impress. It can be made with various mushrooms, wild or tame. In this case, the wonderful flavour of hedgehog mushrooms shines through to become a highlight of the dish. Be sure to use heavy whipping cream for things to set well. However you make it, don't worry. You won't be asked to sing for your supper!

**This recipe is calibrated for a relatively small batch of hedgehog mushrooms. For a larger batch of mushrooms, use the ingredient amounts in parentheses.**

| | |
|---|---|
| 1tbsp | Butter |
| ¼ cup | Bread crumbs, Panko |
| 2 (4) tbsp | Butter |
| ½ (1) | Onion, small, diced |
| 1½ (3) cups | Hedgehog mushrooms, (*Hydnum* sp.) |
| ½ (1) tbsp | Lemon juice |
| To taste | Salt and pepper |
| 1½ (3) tbsp | Flour |
| 1 (2) cup(s) | Whipping cream (33%) |
| 2 (4) | Egg yolks, beaten |

Preheat oven to 350°F. Melt 1 tbsp butter in a skillet (med. heat) and toast breadcrumbs while stirring until they are golden-brown. Remove from skillet. Melt ¼ cup butter in skillet and sauté onion until limp and translucent. Meanwhile, heat the cream in a sauce pan until scalding (but not simmering or boiling). Add mushrooms to skillet and cook until limp and browned (ca. 5-6 min.). Stir in the lemon juice and season with salt and pepper. Rapidly stir flour into the skillet to prevent clumping, then stir in about half of the cream. When everything is thick, rapidly stir in the egg yolks a little at a time, then the remaining cream. Cook until the sauce becomes thick like porridge, then remove from heat and pour into a casserole or baking dish. Sprinkle toasted breadcrumbs on top and bake 25-30 min. until the surface is lightly browned and bubbly. Remove from oven and let stand 10-15 min. to set (but serve while still warm). Serves 2-3.

---

1    Ami. 2005; Butts, V. 2024.

# Duck à l'orange w/ wild rice & hedgehog mushrooms

*Fungi* magazine (published and edited by my colleague Britt Bunyard) features (among other things) articles on gourmet cooking with mushrooms. When an image of duck confit and hedgehog mushrooms appeared on the magazine's Facebook page[1], I knew it was time to go on a recipe adventure. The notes mentioned a sauce made with peppers and oranges—essentially this was a spicy duck *à l'orange*.[2] Derived from an earlier dish called duck bigarade, duck *à l'orange* is a French dish that was popularized by Julia Child in the early sixties.[3] I fell in love with duck *à l'orange* while spending a few weeks in Montpellier, France. I must have had it there at least three or four times. This version includes wild rice, also known as *manoomin*, 'Canada rice,' or 'water oats.' Four wild rice species are found in the genus *Zizania* (regular rice is found in *Oryza*) The rice mix used here is a savory companion for the hedgehogs as well as the duck and orange sauce.

## Duck & sauce

| | |
|---|---|
| 1 | Duck, 2.7 kg, frozen |
| Some | Salt & black pepper |
| Some | Paprika |
| 1 cup | Water |
| 2 tbsp | Sunflower oil |
| 2 | Red carrots, chopped |
| 2 | Tomatoes, chopped |
| 2 stalks | Celery, chopped |
| 4 | Leeks, small,[4] chopped |
| 1 | Onion, chopped |
| 2 cloves | Garlic, minced |
| 2 | Bay leaves |
| 1 tsp | Thyme, dried |
| 3 tbsp | Flour |
| 2 tbsp | Tomato paste |
| 2 cups | Chicken stock |
| 3 tbsp | Red Wine[5] |
| 3 tbsp | Sugar |
| ¼ cup | Vinegar, red wine |
| 4-5 | Oranges |
| 2 tbsp | Grand Marnier® |
| 2 tsp | Blackcurrant jelly |
| 1 tbsp | Butter |

## Rice & mushrooms

| | |
|---|---|
| 1¼ cups | Brown rice, long-grain[6] |
| ¾ cup | Wild rice |
| 1 tbsp | Dried parsley |
| 2 tsp | Italian seasoning[7] |
| 3¼ cups | Chicken stock |
| 1 cup | Hedgehog mushrooms (*Hydnum* sp.), sliced |
| ½ | Onion, chopped |
| 2 cloves | Garlic, minced |
| To taste | White pepper |

---

1     Spahr, D. 2024.
2     Pépin, J. 2023.
3     Shaw, H. 2017.
4     Smaller leeks provide a more tidy look for the sauce, since they can simply be sliced rather than chopped up.
5     Côtes du Rhône
6     Use long-grain brown rice. Its cooking time more closely matches that of wild rice. White rice will become mushy or burn if cooked as long as wild rice.
7     (Or *herbs de Provence*)

**Roasting the duck:** Thaw the duck and then preheat oven to 350°F. Remove (and conserve) the duck wings.[1] Also remove (and conserve) the neck and giblets from the internal cavity. Sprinkle salt, pepper, and paprika over the surface and rub it into the skin and inside the cavity. Roast the duck uncovered in a roasting pan, ca. 1 hour 6 min. per kg (equivalent to 30 min. per pound). A typical bird of ca. 2.7 kg will require almost 3 hours. The duck should be roasted so that the internal temperature (measured with a meat thermometer at the thigh joint) reaches at least 185°F (85°C).

**Preparing the sauce:** While the duck is roasting, heat oil in a large saucepan or small pot and sear the conserved wings, neck, and giblets (ca. 10 min., med.-high heat). Remove the duck parts once they are browned[2] and add carrots, tomatoes, celery, leek, onion, garlic, bay leaves, and thyme and cook until vegetables are softened (ca. 5 min.). Stir in flour and tomato paste, then gradually add 2 cups of stock and the wine. Reduce heat and simmer for an hour. In a third saucepan, bring vinegar and sugar to a boil and simmer (med.-high heat) ca. 4 min. (until syrup turns a rich caramel colour). Strain the pulp from the juice of 2-3 oranges (ca. 1 cup), and stir it into the sauce along with zest from the other two oranges (reserve the fruit sections). Add Grand Marnier® and blackcurrant jelly. Simmer a few more minutes, stir in butter and season with salt and pepper. After simmering a few more minutes, stir this into the sauce in the first sauce pan. Let everything simmer a few more minutes, then lower heat and cover to keep warm.

**Cooking the rice:** Bring stock for the rices to a boil in a sauce pan. Blend rices in a bowl with parsley and *herbes de Provence*. Add the mix to boiling stock, reduce heat to a simmer, mix in the mushrooms, onion, and garlic, season with salt and white pepper, and cover. Simmer 50 min. and let stand 5 min. before fluffing the rice with a fork. Cover to keep warm and set aside.

**Finishing:** When roasted, remove duck from oven and lift from the roasting pan to let juices drain. A roasting rack will do nicely, or you can lift it onto a wire rack placed over another roasting pan. Stir ½ cup of the juices into the orange sauce and fish out the bay leaves if you can spot them. Carve and plate slices of duck alongside orange sections on a bed of the mixed rice and mushrooms. Pour the orange sauce over top and serve warm. Serves 6-7.

---

1    Some recipes say the skin should be punctured with the tip of a knife (at an angle so that the meat isn't pierced) to allow fat to drain during roasting. Store-bought duck in B.C. will likely be a lean Asian duckling without need of this step.
2    The wings, neck, and giblets can be boiled along with the leftover carcass (once the desired meat is cut away) to make a stock broth.

# Pork tenderloin w/ mushroom sauce[1]

This dish wasn't part of the original plan for this book. A pork tenderloin was purchased to test a recipe for matsutake (found in Part 3 of this series). Of course, you can't buy just a little bit of tenderloin. The entire package weighed well over a kilogram. Not wanting to waste this very nice cut, it had to be put to good use and shared between recipes. After a quick check, it seemed that there were more recipes involving pork tenderloin and wild or domesticated mushrooms than you could shake a stick at (or in my case, a ballpoint pen). There on the internet search page were recipes from various regions with creminis, shiitakes, chanterelles, porcini, truffles....[2] Why not Canadian hedgehogs? In this dish, the hedgehog mushrooms play more of a supporting role to other mushrooms, sharing the spotlight with creminis, porcini, and oyster mushrooms. Have no fear, the end result is a scrumptious winter feast perfect for nice, quiet dinner by the fireplace. Don't leave out the applesauce, by the way. It makes all the difference here.

| | |
|---|---|
| 700 g | Pork tenderloin[3] |
| 4 tbsp | Flour, all-purpose |
| ½ tsp | Seasoned mushroom powder |
| ¼ tsp | Salt |
| ⅛ tsp | Pepper, fresh ground |
| 3 tbsp | Butter |
| ⅓ cup | Hedgehog mushrooms (*Hydnum* sp.), sliced |
| ⅓ cup | Oyster mushrooms (*P. ostreatus*), sliced |
| ⅓ cup | King boletes (*B. edulis*), sliced |
| ⅓ cup | Creminis (*A. bisporus*), sliced |
| 1 | Shallot, chopped |
| 3 tbsp | Dry white wine (Riesling) |
| ½ cup | Mushroom broth |
| 3 tbsp | Apple sauce |
| 1 tsp | Sage, dried |
| ½ cup | Table cream |
| Some | Chives |

Slice the tenderloin at a brief angle into medallions and flatten them a bit by laying the flat of a large knife on top of them and lightly pounding.[4] Mix the flour, seasoned mushroom powder, salt, and pepper on a plate and drag the medallions through the flour mix to coat them (conserve the leftover flour). Melt 1 tbsp butter in a skillet and sauté the pork until it is lightly browned (2-3 min. per side). Transfer to a plate and keep warm. Melt remaining 2 tbsp butter and sauté the shallot until it is soft and translucent (2-3 min.). Add the mushrooms and continue to cook 5-6 min. until they are browned and limp. Deglaze with the wine and stir in the mushroom broth and applesauce. Cook until the liquids start to reduce (ca. 3-5 min.). Add sage and adjust the other seasonings. Stir in cream and simmer 2-3 min. Stir in conserved seasoned flour and simmer and stir occasionally until the sauce reaches the desired thickness. Plate the pork medallions, pour the sauce on top, and garnish with chives. Serves 3-4.

---

1   It was very tempting to call this dish 'Porky porcupines,' but I couldn't bring myself to do it. Porcupines, after all, are different animals compared to hedgehogs (and neither one is a mushroom).

2   E.g. Maclean, F. 2012.; Parisi, G., 2014; Bell, G. 2021; Duckhorn Wine Co. 2024; S., Lucille (threeovens). 2024.

3   The meat you purchase may come in much larger amounts. This being the case, you might cut everything into medallions, use the appropriate amount for this recipe, and freeze the rest for later use. There is another recipe for pork tenderloin with mushroom (*Agaricus bisporus*) and truffle sauce on p. 68.

4   Some recipes have you stuff the meat. One slits the meat lengthwise, folds it out 'butterfly' fashion, and pounds it to make it thinner and even. One then lays sauce on top, rolls up the meat, ties it off with strings, and bakes it, testing with a thermometer. We're using a much simpler approach here, but if you're into that sort of thing, check out the recipe for truffled chicken roulade on p. 63.

# Swedish meatballs w/ mushroom sauce

As Vancouver Island's mild winter begins, chanterelles disappear from the forest while hedgehog mushrooms continue to fruit. Hedgehog flavour resembles that of chanterelles—chefs can make a facile transition from one to the other without changing menus. That's what we're doing here, with Swedish meatballs. There are lots of recipes for Swedish meatballs—some sweet, some savory, some using reindeer or 'elk' (moose) and some with chanterelles.[1] This version uses ground venison and a savory, creamy mushroom sauce[2] made with hedgehog mushrooms, Icelandic *skyr* (see p. 24) and a bit of lingonberry sauce to give it some sweetness and colour. You can always use alternative ingredients. But if you're going to all of the trouble to make Swedish meatballs, using the listed ingredients will certainly impress.

## Meatballs

| | |
|---|---|
| 3 slices | Bread w/o crusts |
| ½ cup | Milk |
| 2 lbs | Ground venison |
| 2 | Eggs |
| 1½ tsp | Salt |
| ¾ tsp | Pepper |
| 1½ tsp | Allspice |
| ¾ tsp | Caraway seeds |
| 1 | Onion, diced |
| ¾ cup | Flour |
| ½ cup | Sunflower oil |
| 3 tbsp | Butter |

## Sauce

| | |
|---|---|
| 1 | Onion, diced |
| 1 cup | Hedgehog mushrooms (*Hydnum* sp.), sliced, dried |
| 2 cups | Beef stock |
| ¼ cup | Skyr (see p. 20) |
| ¼ cup | Milk, 3½% |
| ¼ cup | Lingonberry sauce[3] |

Rehydrate hedgehogs in warm water (ca. 45 min.) Mix milk with bread in a bowl, using a fork to mash it into a rough paste. Add eggs, salt, pepper, allspice, caraway seeds, onion, and venison. Mix gently (without squeezing or kneading) until everything is combined. Use a small spoon to scoop out mix. Sticky venison will require gentle handling to form it into small balls. Coat meatballs with flour and place on a baking tray covered with wax paper. Heat oil in a skillet (med. heat) and then melt butter in it (oil should come rise to half the height of the meatballs). Fry meatballs on one side (5 min.), then turn to cook other side 5 min. Remove from the skillet and set aside (keep warm). For the sauce, pour all but 2-3 tbsp of oils from skillet into a sauce pan. Sauté onion in the skillet (med.-high heat) until soft and translucent (2-3 min.), then dice and add the mushrooms, cooking until limp and browned (4-5 min.). Deglaze with sherry, scraping any fond from the bottom of the pan. Add ½ cup oil from sauce pan and stir in 2-3 tbsp of the remaining flour to make a roux. After ca. 1 min. slowly stir in stock. Cook and stir until sauce thickens (1-2 min.). Adjust seasonings (salt and pepper). Mix skyr with milk and lingonberry sauce in a bowl. Add the meatballs and cook ca. 1-2 min. to allow flavours to permeate the meat. Serve with mashed potatoes, green beans, and lingonberry sauce. Serves 5-6.

---

1    Shaw, H. 2010; Butler, A. 2022; Totally Swedish. 2024.
2    Bergo, A. 2014c.
3    Alternatively, cranberry sauce.

# Enokitake

Cultivated enokitake (*Flammulina filiformis*) growing comfortably in their plastic jars at a production plant in Taiwan.[1]

---

1    Photograph by R. Winder. This is not a field guide.

# Biology & Ecology

**Enokitake :** *Flammulina filiformis* is a cultivated (and widely exported) mushroom from Eastern Asia. In Japan, this species is famously known as Enokitake ('*Enoki*' = Japanese hackberry tree, *Celtis sinensis* var. *japonica*, and '*take*' = mushroom). Other common names for this fungus include: 'lily mushroom' 'golden needle mushroom' ('*jin zhen gu*') or 'gold mushroom' ('*jingu*') or '*dongu*' in China , '*futu* mushroom' (India), '*paengi beoseot*' (Korea), '*kim cham*' or '*tram vang*' (Vietnam). There is a cultivated variety of *F. filiformis* called 'golden enokitake' that has darker, golden-brown coloured caps when compared to the usual pale white colour of the commonly grown varieties. Until recently, *F. filiformis* was called *F. velutipes*, a very similar mushroom that grows in temperate areas around the northern hemisphere. However, a recent genetic study and analysis of phylogeny concluded that cultivated forms in Asia and those most commonly found in B.C. and the Pacific Northwest are actually *F. filiformis*.[1] Common names for *F. velutipes* include 'velvet foot' 'velvet shank', 'velvet stem', 'winter mushroom', 'winter ear' or '*téli fülőke*' (Hungary).

**Habitat:** *F. filiformis* is saprophyte, living on stumps, logs, and living wood of hardwood trees, including Japanese hackberry. In B.C. and the Pacific Northwest, *F. velutipes* occurs from September through March, but principally fruits in clusters during the reduced light of winter. Unlike *F. filiformis* it can sometimes be found on poplar.

**Cultivation:** Modern production methods prefer cultivation of the mushroom in bottles with sterile media,[2] often in chambers with controlled temperature, humidity, light, and other factors. Effectively, it is a way to turn woody materials into food. Cultivated forms look a bit different than the dark-brown wild type; they are usually pale (or sometimes golden brown) and more tall and slender with diminutive caps.

**Chemical defences:** During fruiting, *F. filiformis* produces flammutoxin,[3] a cytolytic (cell-destroying) polypeptide similar to phallolysin, a cytotoxin produced by the death cap (*Amanita phalloides*).[4] Studies have shown that flammutoxin has antiviral properties in plant systems.[5] *F. filiformis* also produces variety of volatile chemical compounds (terpenes) that may play a role in attracting mites that would help to spread spores.[6]

Turning sawdust into food. Culture bottles with mycelium of *F. filiformis* being inspected as they zip along a conveyor track between controlled environments at a production plant in Taiwan. Photo by R. Winder.

---

1    Wang, P. *et al.* 2018.
2    Stamets, P. 2005.
3    Lin, J., *et al.* 1975.
4    Bernheimer, A. and Oppenheim, J. 1987.
5    Wu, L., *et al.* 2017.
6    Li, H., *et al.* 2018.

# Consumption & nutrition

**History:** Cultivation of *F. filiformis* started as early as the 8th Century in China[1] and has been practiced for three hundred years in Japan, first on wood and later on sawdust and rice bran.[2] Enokitake are currently among of the most widely consumed mushrooms (with the current world market hovering around $US 1.5B), and well worth adding to any arsenal of mushroom recipes.

**Nutrients & trace elements:** The nutritional profile of *F. filiformis* is shown in Table 1 (p. 4) and trace elements in Table 2 (p. 5).[3] As the fungus can thrive on contaminated media, caution should be used regarding the types of cultivation substrates or harvest sites.

**Aroma, flavour & quality:** The flavour of enokitake is variously described as mild, delicate, fruity, savory, and nutty. Volatile compounds in the mushrooms mainly consist of alcohols, aldehydes, and ketones, with 3-octanone predominating. Drying increases the alcohol, acid, and ester content while decreasing ketones.[4] The presence of saccharide compound known as D-arabinitol in the cultivated form has been proposed as an indicator for quality control during production.[5]

**Bioactivity:** Bioactivity detected in enokitake is summarized in Table 3 (p. 6). Experiments using heat treatment of *F. filiformis* extracts (60°C, 5 min.) have shown that FIP-5, a potent immune (T-cell) activator,[6] persists after heating while the flammutoxin is inactivated.[7] One paper found that FIP-5 has anti-inflammatory properties that counter asthma reactions in mice.[8] Moreover, fibre content in this mushroom was effective in lowering total serum cholesterol levels in one rat study[9]—the presence of fibre, polysaccharides, mycosterols, and relatively low sodium levels may also help to lower blood pressure.[10] Other studies have focused on evidence of potential anti-cancer effects of immune-system modulating proteins in the mushroom, both in demographic studies of *F. filiformis* farmers in the Nagano region of Japan, and in laboratory experiments focused on breast, prostate, liver, and other cancers.[11]

*F. filiformis* is known to produce flammutoxin, a cytotoxin that destroys red blood cells. Luckily, the compound is heat labile[12] (broken down by heat).[13] A laboratory study also found that *F. filiformis* can increase creatine kinase activity in mice. While this activity has been associated with rhabdomyolosis (destruction of striated muscle) in other mushrooms, further investigation is needed.[14] It would probably be smart not to overdo consumption until more is known.

**Lookalikes & safety:** Given the possibility of confusing *F. velutipes* with other mushrooms, and also the increasing availability of cultivated *F. filiformis* in grocery stores, it would be advisable to focus on the commercial product rather than a wild harvest. For example, in B.C. and the Pacific Northwest it's possible that a beginner could mistake *Galerina marginata* (a.k.a. 'Funeral Bell') for *F. filiformis* or *F. velutipes*. *G. marginata* and other members of the genus have the same amatoxin as the deadly 'Death Cap' (*Amanita phalloides*). So yes, those monikers should give one pause. A mistake could have dire consequences—even experienced folk tend to head to the grocery store for this one.

All of the above being said, the commercial product isn't totally problem-free. There have been a few recalls of enokitake mushrooms imported to Canada. Contamination with bacteria (*Listeria* sp.) is sometimes mentioned. Consumers should be aware of any outstanding recall notices when purchasing this mushroom in stores.

---

1    Yang, X. 1986; Wang, N. 1995.
2    Nakamura, K. 1981.
3    Brunnert, H., and Zadražil, F. 1983; Zhu, C. *et al.*, 2014; Siwulski, M., *et al.* 2019.
4    Yang, W., *et al.* 2016; Fang, D., *et al.* 2017.
5    Cai, H.*et al.* 2013.
6    Wang, P., *et al.* 2004.
7    Tung, C., *et al.* 2018.
8    Chu, P., *et al.* 2017.
9    Fukushima, M., *et al.* 2001.
10   Yeh, M., *et al.* 2014.
11   Powell, M. 2014.
12   Watanabe, A., *et al.* 2004.
13   So you are going to **cook** this mushroom and not eat it raw, right?
14   (In humans, one would assume.) See: Mustonen, A., *et al.* 2018.

# Recipes

## Sautéed enokitake

Like many choice mushrooms, enokitake are versatile. That's why so many grocery stores are starting to carry them! They are a favourite *ad hoc* addition to vegetable stir-fry dishes and similar fare, often added towards the end of cooking due to their relatively thin shape and smaller size. Nevertheless, they have a nicely chewy texture and offer a slightly sweet, faintly peppery taste. The ingredients for this recipe are relatively simple as the enokitake flavor takes centre stage, but if you start with this one, you'll learn more about how you want to use this mushroom in other cooking. This recipe is adapted from several online sources.[1]

| | |
|---|---|
| 200g | Enokitake (*Flammulina filiformis*) |
| 2 tbsp | Sesame oil |
| 1 clove | Garlic, minced |
| 1-2 tbsp | Soy sauce |
| 1 | Scallion or green onion, finely chopped |

In a wok or similar pan, heat the oil on medium heat, add the minced garlic and some of the chopped scallion or green onion and cook until it is golden-coloured. Rinse mushrooms in cold water, cut off the woody base, and separate. Add the mushrooms and toss while cooking to coat them (a few min. until they are limp). Add the remaining scallion or onion and soy or stir-fry sauce and cook another minute or so. Serves 1-2 (3-4 if used as a topping over rice or noodles).

## Nametake

Nametake (in Kanji, 'slippery' + 'mushroom,' a.k.a. enokitake) is a popular side-dish in Japan. Soy sauce and *mirin* (a sweet, savoury rice wine) are traditionally featured.[2] Mirin is made with a glutinous rice, distilled alcohol, and rice cultured with *koji* (*Aspergillus oryzae*, also used to ferment, soy sauce, sake, etc.). Slightly acidic and low in alcohol, mirin balances saltiness and provides umami taste in many dishes (in teriyaki sauce, tofu broths, etc.).[3] Mirin seasonings are available through various online sources if not in your local store--alternatives include dry sherry, sweet marsala wine, rice cooking wine, sake, dry white wine, and rice vinegar (mix in about ½ tsp sugar per tbsp of the alternate ingredient).[4]

| | |
|---|---|
| 200g | Enokitake (*Flammulina filiformis*) |
| 1-2 tbsp | Soy sauce |
| 3 tbsp | Mirin or sweet white wine |

Cut off the woody base of the mushrooms and separate into three bundles. Add the soy sauce and wine to a wok or similar pan. Cook over medium heat several min. until the liquid steeps (and alcohol evaporates). Lower heat to simmer and add soy sauce and sugar, stir, then add mushrooms. Cook until the mushrooms are limp and brown from the sauce. Serves 1-2 as a simple side dish or 3-4 served as a topping on white rice or tofu, sandwiches, noodles, etc. It's fine warm or cold. Stored and refrigerated in a heat-treated, airtight container, nametake can keep for several days.

1    Brian. 2015; Overhiser, S. and Overhiser, A. 2021; Judy. 2017.
2    Jacobson, A. 2016; Slutsky, A. 2021; Uncut Recipes. 2021.
3    Smith, K. 2021.
4    Cebulah, T., and Wicks, L. 2016.

# Enokitake & ramen soup

This recipe is modified from a more elegant recipe calling for somen noodles or vermicelli, *dashi* or light vegetable stock, and the addition of *arame* or *hijiki* dried seaweed.[1] The focus here is more on the texture of the mushroom in contrast with the other ingredients (the mushrooms are hard to see when they blend in with the noodles). In this recipe, ramen noodles are quick to prepare and easy to find prepackaged with powdered vegetable flavouring. These wheat-based noodles were first brought to Japan by Chinese immigrants in the late 19th Century. Demand greatly increased at the end of WWII, when rice harvests failed and Japan lost agriculture (including rice) from former colonies in China and Taiwan. During the occupation period, the Americans imported wheat to deal with food shortages, and some of it found its way into ramen. Slow government food distribution and official rationing in the postwar period led to a flourishing and increasingly essential black market, where ramen was widely available. Ramen continued to increase in popularity when rationing ended, and ramen soups started to feature soy sauce, pork, chicken, and *niboshi* (dried sardines). This sort of food with strong flavours and fats was known as a 'stamina food.'[2] Ramen noodles eventually became one of Japan's most popular foods—Tokyo alone is thought to have over 5,000 ramen shops. Regarding the walnut oil used in this recipe, if your local grocer doesn't have it, you might try looking in a health food store. Here's hoping this soup will not only satisfy your palette, but fortify and nourish you!

| | |
|---|---|
| 1 cup | Enokitake (*Flammulina filiformis*) |
| 1 pack | Ramen noodles (86 g) w/ veg. flavour packet |
| 1 | Carrot, large, sliced |
| 1½ tsp | Lemon juice |
| 1½ tsp | Walnut oil |
| 1 | Green onion top, sliced |
| 1 dash | Seasoned mushroom powder[3] |

Use a potato peeler to make thin lengthwise carrot slices, chop them, and set them aside. Chop the woody base off the mushrooms, separate them, and set aside. Bring water (w/ flavour packet contents) to a boil and reduce heat to simmer. Add chopped carrot and noodles, cover, and cook 3-4 min. (removing the cover to stir occasionally and separate the noodles – typically, the noodles are ready when the soup starts to bubble up to the top of the pan). Stir in vegetable flavouring from mini-packages, lemon juice, and oil, add the mushrooms, and cook another few min. until mushrooms are limp. Pour into shallow soup bowls, season with mushroom powder, and garnish with onion slices. Serves 2-3.

---

1    Jordan, P. and Wheeler, S. 1995.
2    Lu, H. 2018.
3    Suggested recipe in Part 1.

# Winter truffles

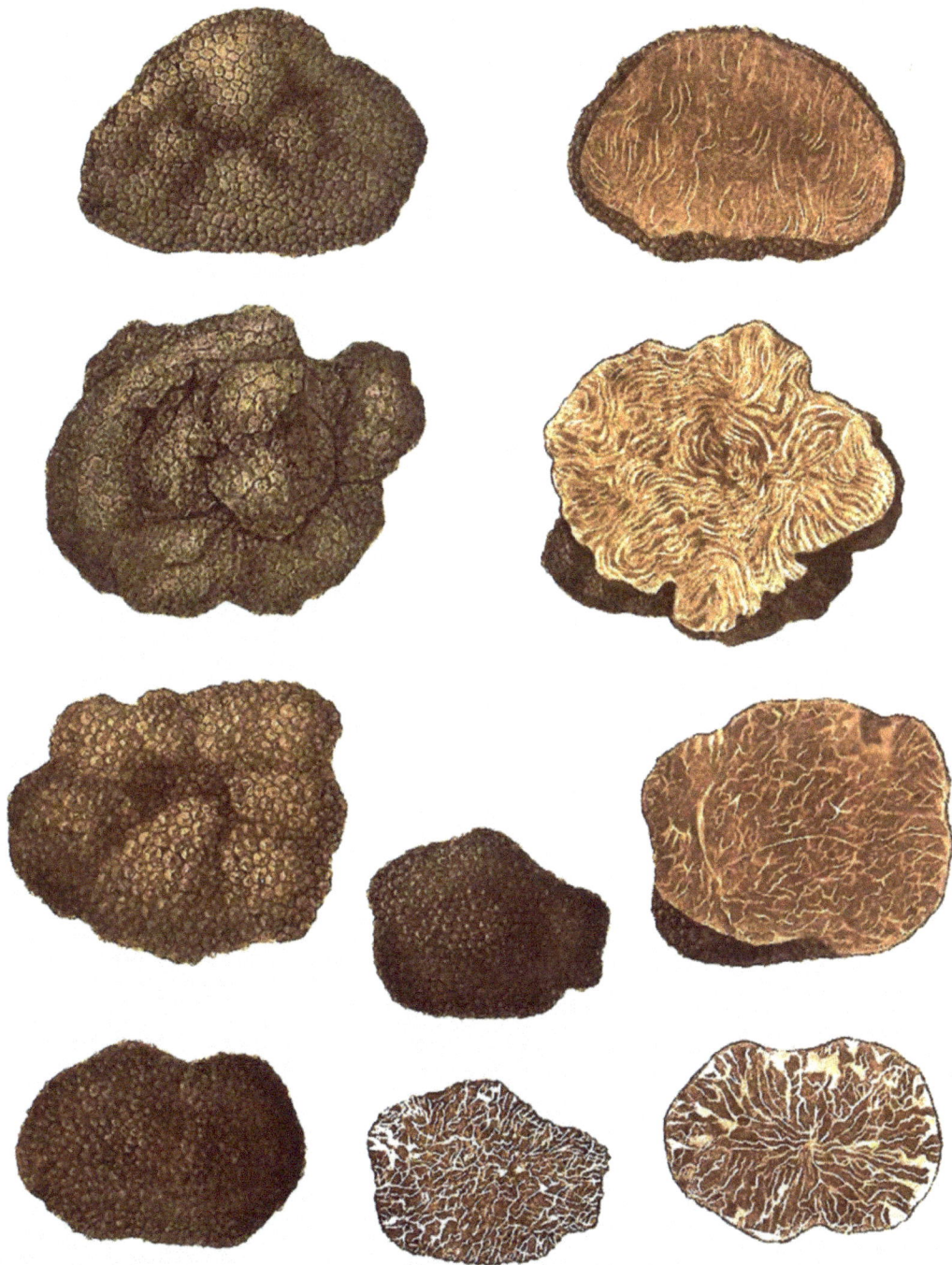

Edible truffles depicted in Roques, J., 1832 as '*Tuber cibarium.*' *T. cibarium* is now an illegitimate name for various black truffle species—these are all likely portrayals of *T. melanosporum*. The truffles at top are "*Truffe du Périgord.*" The pair below that are "*T. du Quercy.*" Below that at left and right are "*T. de la Drôme*" and "*T. de Vaucluse.*" At bottom centre, the smaller pair are "*T. du Gard.*"[1]

---

# Biology & Ecology

**Winter truffles:** Last but certainly not least, we finally come to the winter truffles of Eurasia. Similar to the previous truffle chapters in this series, the species in this chapter might actually appear earlier or later than winter depending on locale, weather, etc. Nevertheless, they are predominantly found during some part of the winter. *Tuber melanosporum* (a.k.a. Périgord truffle, black truffle, French black truffle, black diamond) is currently the most widely sold truffle. Covered with dark pyramidal warts and with its dark spores and unique aroma, it is probably the species most people are familiar with when someone mentions truffles. It was formally described by Italian naturalist Carlo Vittadini in 1831. *T. brumale* (a.k.a. Muscat truffle, winter truffle, black winter truffle) can also occur in the same orchards as *T. melanosporum* (where it is unwelcome and can also be confused with it). Regarded as the 'lesser brother' of the Périgord truffle, it was also described by Vittadini in 1831 and enjoys about half the value of its more famous sibling. *T. indicum* (a.k.a. Himalayan black truffle, Asian black truffle, Chinese black truffle, Asian winter black truffle) is found in the Himalayan regions of India, Nepal, Bhutan, China and also in Japan. Like *T. brumale* it both resembles *T. melanosporum* and is considered commercially less desirable, something that has created issues in the past when supplies of the Périgord truffle were low and counterfeits in the form of *T. indicum* appeared in the market to take advantage of the situation.

*T. borchii* (a.k.a. the bianchetto truffle, Marzuolo truffle, whitish truffle, white summer truffle, Italian spring white truffle, Tuscany white truffle) is a white truffle regarded as the less famous cousin of *T. magnatum*. Described by Vittadini in 1780, it can be found in the same habitat as *T. melanosporum* but has buff-coloured flesh that eventually develops reddish-ochre shades and with white veins eventually turning brown.

**Habitat:** All of the truffles in this section are mycorrhizal. *T. melanosporum* and *T. brumale* both thrive in calcareous soils of southern Europe (France, Spain, Italy, and Croatia), share similar hosts, and are harvested from November to March in their native range. Hosts associated with *T. melanosporum* include common oak (*Quercus robur*), holm oak (*Q. ilex*), hazel (*Corylus* spp.), cherry (*Prunus* spp.), and other deciduous trees. While also liking oak hosts, *T. brumale* has a preference for hazels and lindens (*Tilia* spp.). Relative to *T. melanosporum*, *T. brumale* can be found in similar or less calcareous soils and similar or more humid conditions. Described from specimens collected in India in 1989, *T. indicum* matures in winter and associates with conifers such as Yunnan pine (*Pinus yunnanensis*), Chinese white pine (*P. armandii*), and Evelyn keteleeria (*Keteleeria evelyniana*). Oaks are also reported hosts in India. *T. indicum* likes calcareous soils, growing at altitudes ranging between 1800m -3000 m above sea level. It was recently introduced to North America (in Oregon) and Italy, and has been shown to have the capacity to form mycorrhizae with North American trees (loblolly pine—*P. taeda* and pecan—*Carya illinoinensis*). Whether it has the potential to be an invasive threat to established truffle orchards is still unknown.[1]

*T. borchii* grows throughout Europe in association with oaks and (to a certain degree) conifers. It produces fruiting bodies from December to June. Like the other truffles it can grow in soils with relatively high pH (7-8), but it can also be found at lower pH (6-7, in some cases as low as 5.2).

**Biology of the *brulé*:** Truffles often form a clear zone around their mycorrhizal host tree called a brulé (in French, burned area). The zone isn't literally burnt, it just figuratively resembles an area where fire might have cleared away vegetation at some point. While it may look like a dead zone, there's a lot going on in a brulé. The fungus is forming a symbiotic association with its host tree and is simultaneously reducing the presence of nearby plants. It is also modifying the community of soil bacteria and fungi in the brulé while affecting the physiology of its host. Let's look at some details, using *T. melanosporum* as an example.

Recent research has outlined a three-way interaction between *T. melanosporum*, holm oak host trees, and nearby companion plants. When the ascomycete forms ectomycorrhizae with the roots of its tree host, it also interacts with nearby plants. Abundance of the mycelium increases and ectomycorrhizae on the roots of companion plants decreases (along with germination of associated seeds).[2] This is accompanied by a decline in the growth and nutrient status of the companion plants while the nutrient status of the host tree benefits. Earlier studies pointed to the association of *T. melanosporum* with necrosis in non-host roots (as a parasite), while more recent work has indicated that *T. melanosporum* (as well as *T. aestivum*) can colonize non-host plants as an endophyte.[3]

---

1    Bonito, G., *et al.* 2011.
2    Taschen, E., *et al.* 2020
3    Plattner, I., and Hall, I. 1995; Schneider-Maunoury, L., *et al.* 2019a.

Amendment of truffière soils with calcium carbonate (lime) is known to increase the production of fruiting bodies. This was confirmed in a study of *T. melanosporum* truffières contaminated with *T. brumale*. The treatment did not eliminate *T. brumale* as hoped but increased the number of truffles produced by both species. Given that other studies find that carbonates are at higher levels in brulé soils vs. outside, one theory is that the truffle fungi acidify the soil within a brulé, freeing carbonates that can cause nutrient deficiencies that in turn lead to chlorosis (loss of chlorophyll) and decline of non-host plants. This in turn favours growth of the fungus, formation of truffle mycorrhizae, and development of the brulé.[1]

While *T. melanosporum* is managing vegetation, it is also having its way with other soil microbes within the brulé. The diversity of ectomycorrhizal fungi and basidiomycetes in the soil of brulés in oak plantations tends to decrease while mycorrhizal helper bacteria and other fungi remain associated.[2] In a study of oak truffières producing *T. melanosporum* in Spain, bacteria associated with the truffles included *Singulisphaera limicola*, *Nannocistis excedens* and *Sporosarcina globispora*, while associated fungi included *Mortierella* and *Cryptococcus* species.[3] A study of galcham oak (*Q. aliena*) seedlings inoculated with *T. melanosporum* and *T. indicum* found that richness of denitrifying bacteria (with the nir-K gene) was greatest in soils associated with the *T. indicum* plants and lowest in controls. The same study found some effects on host seedlings: peroxidase enzyme and nitrate levels in inoculated seedlings were elevated and phosphorus levels decreased. Growth rate, on the other hand, wasn't really affected.[4] Nitrogen-cycling events may also be happening within the fruiting bodies that are found in brulés, not just in the soil or truffle mycelium. Researchers using high-throughput DNA sequencing to look at *T. indicum* from nine areas in China found that certain bacteria (*Bradyrhizobium* spp.) were dominant members of the microbial community occurring within the tubers (but not the soil or mycelium). *Bradyrhizobium* ssp. have the potential to be nitrogen-fixers, and the researchers confirmed that at least one of the bacterial strains was a capable nitrogen-fixer.[5]

Truffle fruiting bodies fruiting with a brulé also provide habitat for truffle flies (*mouche de la truffe*, e.g. *Helomyza tuberivora*). When they probe in the earth for the precious 'black diamonds, truffle hunters also look for the presence of red-coloured truffle beetles (*Leiodes cinnamomea*) or also brown-coloured earthworms. Both are considered sure signs that they are getting close to their quarry.[6]

**Cultivation:** Given the high prices and increasing demand that even second-rate truffles command in the marketplace, no one should be surprised that there is ongoing interest in developing and improving cultivation methods for truffles. This is especially the case where habitat has been in decline due to various factors (climate change, natural forest succession, human activity and changes in land use). Recent declines in truffle productivity have promoted increased interest in cultivation to maintain the market. It is estimated, for example, that about 40% of *T. melanosporum* production now comes from Spanish plantations.[7]

The mycelium of truffle species can be cultured on agar media or in liquid media, using substrates such as Murashige and Skoog Basal Medium.[8]

The general approach to cultivating truffles involves inoculating the roots of host tree seedlings with truffle spores or mycelium, testing to verify that the correct mycorrhizae have indeed colonized the seedling roots, out-planting of the tree seedlings, and managing the plantations to control weedy plants. Some businesses offer spores, mycelium, or even inoculated tree seedlings for would-be truffle farmers to establish their own *truffière* (truffle orchard or plantation). *T. melanosporum* can be cultivated beyond its native range. Cultivation has been established elsewhere in Europe (Wales), the Southern Hemisphere (Australia, New Zealand, Argentina, Chile, and South Africa), and in North America (North Carolina, Tennessee, Arkansas, Texas, California, Oregon, and British Columbia).

The success and productivity of plantations is determined by the details of site location (including soil type and site preparation), source of inoculum, inoculation method, and orchard establishment and management. Presence and

---

1    García-Montero, L., *et al.* 2009.
2    Napoli, C., *et al.*, 2009; Mello, A., *et al.* 2015.; Taschen, E., *et al.* 2020.; Herrero de Aza, C., *et al.* 2022.
3    Herrero de Aza, C., *et al.* 2022.
4    Kang, Z,. *et al.* 2022.
5    Chen, J., *et al.* 2019.
6    Toussant-Samat, M. 1987.
7    Rodrigues, A. 2008.
8    Shah, N., et al. 2022.

spacing of trees inoculated with different mating strains of the fungus may have an important impact on productivity.[1] When planting the trees, it may be important to maintain buffers to exclude other mycorrhizal fungi that can invade a plantation and displace the truffles. Orchards must have soils with the appropriate pH (relatively high for *T. melanosporum*), and weedy vegetation might need to be kept reasonably in check (keeping in mind that some non-host plants might actually be beneficial).

Trained pigs were once traditionally used to detect truffles during harvests. Pigs do excel at truffle location but unfortunately they are also avid truffle consumers. Truffles are now usually located with the assistance of trained dogs (because they don't consume their quarry). Animals don't find everything—research has shown that the 'spore bank' of managed plantations remains robust, with up to 42% of truffles remaining undetected and unharvested.[2]

---

1    Rubini, A., *et al.* 2014.
2    Schneider-Maunoury, L., *et al.* 2019b.

# Consumption & nutrition

**Truffle history:** Truffles have been consumed since ancient times—much of the information in this section pertains to truffle species in general (before scientific descriptions of species came to be). Although accounts are unclear, truffles may have been part of the diet of Jewish and Sumerian people around 1700-1600 BCE. One anonymous source (ca. 1600 BCE) called them 'mysterious products of the earth.'[1] Other sources list ancient Etruscans, Greeks, Egyptians, and Romans as consumers.[2] There is a theory that the miraculous food *manna* described as saving the Israelites in the Biblical Old Testament might have been desert truffles (*Terfezia* spp.).[3] *Terfezia* spp. grow in drier climates and are milder and less pungent than other types of truffles. The Biblical description mentions that manna was white and tasted like wafers made with honey—a contemporary description of *Terfezia* says that the mushrooms are crisp like a potato and have a nutty flavour tasting of bee's wax.[4]

Meanwhile in ancient China, farmers did not historically consume *T. indicum*, instead feeding it to pigs.[5] It does feature in Traditional Chinese Medicine as a tonic for rejuvenation, restoration of appetite, balanced health, etc. Moreover, an important event on the Indian subcontinent hints that there could have been some human consumption in southern Asia. Sometime around 411-400 BCE, Siddhartha Gautama (the Buddha) was in the city of Pāvā, India (now called Fazilnagar), on his way to Kuśīnagar. A smith named Cunda Kammāraputta offered Siddhartha a meal consisting of rice, cakes, and *sūkaramaddava*. The last item is translated differently depending on the tradition involved. *Sūkara* refers to 'pig' and *maddava* means 'soft,' 'tender,' or 'delicate.' In the Theravada tradition the food may have been tender pig or boar meat, or in the Mahayana tradition it may have been a tender vegetable enjoyed by pigs or boars (i.e., a yam, tuber, mushroom, or truffle). Whatever it was, it was Siddhartha's last meal. He suffered an attack of fatal dysentery, possibly a necrotizing enteritis caused by a bacterial contaminant (*Clostridium*) and aggravated by the high protein diet of some ascetics. Before he died, Siddhartha told his disciple Ānanda to visit Cunda and absolve the smith of any blame or remorse for serving the meal.[6]

Ancient Greeks certainly knew about truffles. In the 5th Century BCE, it was recorded that an Athenian *metic* (resident alien) was granted citizenship in return for preparing an original truffle dish. Greek philosopher Theophrastus (ca. 371-287 BCE, successor of Aristotle) wrote descriptions of truffles in his *Historia plantarum*. He reported that truffles were born from autumn rains or flashes of lightning. A century later, Greek poet Nicander of Colophon supposed that truffles were silt modified by some kind of internal heat. The Greek philosopher Plutarch of Chaeronea (CE 46 - after 119 CE) theorized that lightning formed them in mud.[7] On the other hand, Greek physician Dioscorides (CE 40-90) came closer to the truth, calling them tuberous roots.[8]

The ancient Romans loved truffles, an affection they learned from the Etruscans. Experts believe that the Romans actually consumed *Terfezia* spp.[9] from the many parts of their empire with drier climates. The Roman orator Cicero (106-43 BCE) referred to them as "children of the earth." Cicero and General Pompey could very well have dined on truffles when they visited the fabulously wealthy (and epicurean) General Lucius Licinius Lucullus (he of 'lucullan' banquet and gastronome fame). Truffles often occupied a place of honour at his banquets.[10]

Pliny the Elder (ca. 23-79 CE) later called truffles "callosities of the earth" and "lumps of earthy substance balled together." In his *Naturalis Historia,* Pliny echoed Plutarch's opinion, stating: "When there have been showers in autumn, and frequent thunder-storms, truffles are produced, thunder contributing more particularly to their development." He further stated: "In some places the formation of them is attributed to water; as at Mytilene, for instance, where they are never to be found, it is said, unless the rivers overflow."[11] Pliny wrote that while they appeared after autumn thunderstorms, they usually only lasted for a year and provided the "most delicate" eating in the spring. He also wrote

1   Cascina Alberta. 2023.
2   Mustafa, A., *et al.* 2020.
3   *Ibid.*
4   Allen, K. and Bennett, J. 2021.
5   Specialty Produce. 2023.
6   Wikipedia. 2023.
7   Mustafa, A., *et al.* 2020; Specialty Produce. 2023.
8   Allen, K. and Bennett, J. 2021.
9   *Ibid.*
10  Toussant-Samat, M. 1987.
11  Pliny the Elder. 77 AD.

about an unfortunate Minister of Justice in Cartagena named Lartius Licinius, who cracked his teeth on a *denarius* (coin) in the centre of a truffle when he bit into it.[1] Don't be like Lartius Licinius. Check inside your truffles before you bite into them.

The 1st Century A.D. saw various Roman recipes for truffles included in the *Apicius* cookbook. Some Romans were quite taken with truffles. Take, for example, the case of Titus Alledius Severus. Alledius was a Roman *eques* (member of the upper class originating with Roman knights). His big claim to fame was his support for Aggripina and her scandalous marriage to her uncle, Emperor Claudius (when Claudius asked the Senate to legalize such marriages, Alledius was the only senator to say he also wanted to marry his niece).[2] In his collection of satirical poems, Roman poet Juvenal (ca. 55-138 CE) wrote:

> *"'Tibi habe frumentum' Alledius inquit, 'O Libye, disiunge boues, dum tubera mittas'"* [3]

There are various translations, but the one we'll run with is:

> *"'May your corn be yours,' said Alledius. 'O Libya, untie the oxen while you send the truffles.'"* [4]

On the surface this quote points out how much some elite Romans loved their truffles. Keep in mind, however, that Roman satire had a formal structure, often meant to criticize excesses of the elites and offer moral examples from a somewhat conservative perspective. This quote should probably be taken as having a disapproving tone. Juvenal's *5th Satire* also describes the nasty behaviour of a social climber named Virro. Virro liked to amuse himself by humbling and humiliating his less well-to-do dinner guests—serving those of lower social status a poor menu while he dined on various luxury foods like goose liver, capon, boar, and (if the spring thunder had blessed them) truffles.[5] Don't be like Virro. Share your truffles with your less fortunate dinner guests.

With the fall of the Roman Empire and the start of the Medieval Era, truffle consumption withdrew into obscurity. Between the 5th and 12th Centuries there was not much written about them in the literature of western Christendom. But they were still valued in the Islamic world. During the 6th and 7th Centuries, the Islamic Prophet Muhammad recommended truffles for treatment of eye problems. Avicenna (a.k.a. Ibn Sina) was a 10th Century Islamic physician and philosopher who considered truffles to be a treatment for a variety of illnesses (vomiting, wounds, weakness). Echoing Muhammad, he also prescribed truffle juice for eye inflammations.[6] In the 12th Century, Abd al-Majid ibn Abdun (an Islamic poet of Moorish Iberia) warned against selling truffles near mosques because they were a "delicacy of the dissolute."[7]

Truffles enjoyed a resurgence of interest in western high society during the Renaissance of 14th Century France and Italy. When Charles VI of France married Isabeau of Bavaria in 1385, truffles were served at their wedding banquet (thanks to their introduction by Charles' luxury-loving brother Jean de Berry).[8] One of the leading bores and cranks of the time, Eustache Deschamps, was a soldier and diplomat serving under Charles—he fancied himself a satirical poet, and decided to write a poem about truffles. The commentary comes across less like poetic satire and more like outright contempt. Translated from French:

> *"It is a root of horrible appearance, which one would do well to disguise.*
> *It is black outside, but in decoction it overheats, and the flavour stinks and smells of that.*
> *He who first dug it up was the cause of trouble. This unpleasant plant is taken by mouth.*
> *I have eaten of it, and it left my weary heart in worse case than if I had a tertian fever."* [9]

Don't be like Eustache Deschamps. Don't eat truffles with a weary heart.

---

1    Grigson, J. 1975.
2    Tacitus, Publius Cornelius. Ca. 116 A.D.
3    Roques, J. 1832.
4    Juvenalis, Decimus Junius. ca. 100-127 A.D. Translation provided in part by K. Luther, supplemented by Google Translate.
5    Ibid.
6    Allen, K. and Bennett, J. 2021.
7    Rodrigues, A. 2008.
8    Toussant-Samat, M. 1987.
9    Original quote is from Toussant-Samat, M. 1987. Tertian fever is vivax malaria.

Truffle legends also attribute growing interest in Europe to Catherine de Medici of Florence (1519-1589, spouse of the French King Henry II (1519-1559). Not having been there at the time, yours truly will have to leave the debate over the extent of this influence to others. The Golden Age of Truffles commenced during the 18th Century, with growing French influence in gastronomy.[1] This was further amplified in the 19th Century with the discovery of cultivation practices and Italian efforts to valorise *T. magnatum*. Around this time, the princes of the Savoy region started to present white Italian truffles as diplomatic offerings, increasing the general popularity and prestige of truffles among elites. In 1810, a French peasant named Joseph Talon devised a method of 'cultivating' truffles in the Vaucluse area of Provence. In practice the method consisted of simply planting oak seedlings (without inoculation) in an area where truffles occur (and then waiting about ten years). At any rate, this greatly increased supply. The quantity of truffles sold in Paris during 1828-1829 amounted to 14-17 thousand kilograms, depending on the year.[2] But by the end of 19th Century, over two million kilograms were being sold in France.[3] Current truffle production has waned. The annual market is estimated to be worth about CDN$ 488,000,000 for all truffles. The market share (slice of the pie) occupied by *T. melanosporum* appears to be about 60%. Using an average price of ca. $98.5/oz., that translates to a yearly production of a bit over 84,000 kg for black Périgord truffles.

**The ambience of love:** For better or worse, truffles have a legacy of being regarded as an aphrodisiac. While some writers wax eloquent about truffles creating an ambience conducive to sexual arousal, others are reticent when it comes to considering potential causes. This is generalizing, but some cautious writers are skeptical of any chemical influence. They invoke society's association of truffles with wealth and then link luxurious (decadent?) consumption to piqued interest and a generally increased atmosphere of excitement that might lead to arousal and licentious behaviour. On the other hand, more fearless writers brush that aside and link the musky odour of truffles to smells (and/or pheromones or other properties) associated with sex (and their powers of suggestion). Resolving this debate would likely involve a large-scale scientific study. One would have to compare the degree of sexual arousal experienced by random couples after they consumed truffles versus similar 'controls.' Given that you would not want the random couples to know supposed or potential outcomes, it would be very difficult to slip that kind of experimental design past the ethics departments of most major universities. All we can do at this point is consider the anecdotal evidence and opinions.

In Ancient Greece, Philoxenus of Leucas mentions in his *Symposium* that truffles baked in embers are conducive to 'amorous play.'[4] The Roman poet Juvenal (mentioned previously) took Plutarch's story of lightning-born truffles a step further, speculating that truffles originated from a bolt of lightning hurled by Jupiter at an oak tree.[5] As the father of gods was known for his incredible sexual prowess, and oak trees were sacred to him, and those enticing truffles could be found with oaks.... Hopefully you do see where Juvenal was going with all of this. The Roman physician Galen (Aelius Galenus, 129-ca. 219CE, whose clients included Emperors Marcus Aurelius and Commodus) prescribed truffles to his patients as nourishment and something that would cause a general agitation/excitement that would be predisposing (or conducive) to voluptuousness and sexual pleasure.[6]

Writer Pierre de Bourdeille (a.k.a. Brantôme, ca. 1540-1614) recorded that the amorous *dames galantes* of the 16th Century French court held truffles in high regard, consuming hot pies stuffed with cockerel testicles, artichoke hearts, truffles and other delicacies. These ingredients were supposed to have a 'stimulating effect.'[7]

William Cullen, a lecturer and physician at the University of Edinburgh, said in his *Lectures on the Materia medica* (1753) that truffles were perhaps the only thing that could claim to have authentic aphrodisiac properties. That same century, the very wicked Donatien Alphonse François, Marquis de Sade, is alleged to have used truffles to make his victims more compliant.

Moving to 19th Century, Napoleon Bonaparte I, Emperor of France perhaps did his part to elevate the allure of truffles. Around the year 1810, the story goes that Napoleon conceived his first son, Napoleon-François-Charles-Joseph Bonaparte (*Roi de Rome*) with his second wife Marie-Louise of Habsburg-Lorraine (daughter of the Emperor of Austria), after consuming a truffled turkey. In 1825, French gastronomic author Jean Anthelme Brillat-Savarin discussed the "erotic virtue" of truffles and related the tale of a dinner involving a truffled chicken dish and licentious behaviour

1    Mustafa, A., *et al.* 2020; Allen, K. and Bennett, J. 2021.
2    Moynier, MM. 1836.
3    *Ibid.*
4    Toussant-Samat, M. 1987.
5    Cascina Alberta. 2023.
6    Mustafa, A., *et al.* 2020; Rodrigues, A. 2008.
7    Toussant-Samat, M. 1987.

(there is a pattern beginning to develop here). The hostess had to fend off the uncharacteristic sexual advances of an invited dinner guest (and was also embarrassed that she did not offer a more emphatic rejection). While Brillat-Savarin allowed that the truffles were not exactly an aphrodisiac, he claimed that they were capable of making women more "tender" and men more "likeable" or "apt to love."[1] "He wrote:

> *"Who ever says truffle, pronounces a great word, which awakens erotic and gourmand ideas both in the sex dressed in petticoats and in the bearded portion of humanity"*[2]

Chef Raymond Oliver (1909-1990) claimed that all mushrooms are aphrodisiacs. In a chapter on aphrodisiac cookery in his *The French at Table* (1967), he wrote that the best way to 'employ' truffles was to scrape them, wash them, soak them in cognac, and then eat them raw with a bit of salt. Oliver suggests that 'rosy salt' is the best to use, made by mixing fine salt with paprika and Cayenne. A less direct method suggested by Oliver is to mix oil, pepper, salt, and a bit of wine vinegar in a bowl with some lettuce hearts and raw truffles.[3]

Truffle aromas help to attract creatures that consume them and then spread the spores to other places. It's not surprising, then, that much has been made of the detection of the volatile steroidal compound 5α-androst-16-en-3α-ol (a.k.a. alpha-androstanol) in truffles. It is thought that androstanol can stimulate the female hypothalamus in pheromone-mediated mating behaviours. It has been implicated as a compound affecting female arousal and female mood during menstrual cycles, and combats anxiety and convulsion through modulation of GABA (neurotransmitter) receptors. There are theories that the neurotransmitters dopamine and serotonin can affect arousal (with dopamine in particular improving genital reflexes and copulation instinct). All of this is fine and dandy, but it turns out that the thing that is actually responsible for attracting animals (particularly pigs) to truffles is a relatively unassuming volatile compound named dimethyl sulfide (DMS).[4]

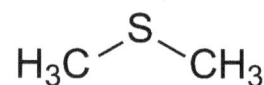

Dimethyl sulfide

There is still a belief that truffles possess an aromatic power of suggestion. The odour of truffles has been described as "the muskiness of a rumpled bed after an afternoon of love in the tropics." Author Elizabeth Luard writes "Not to put too fine a point on it, when ripe and ready, the truffle reeks of sex. You don't get the full impact unless you're there when it's dug." She further notes that they smell "like nothing you'd wish to describe in polite company."[5] Yours truly has sliced open and smelled raw *T. melanosporum* and *T. magnatum*, and cannot vouch for any of these claims—perhaps one should take them with a grain of rosy salt.[6] On the other hand, some friends do claim to have experienced the effect. It's possible that different people detect and perceive the complex profile of truffle scent compounds differently. See the 'Aroma & flavour' section that follows for more details about volatile compounds in *T. melanosporum* and others.

Modes of consumption

If you're reading carefully through this chapter, you've probably realized by now that somewhere along the way there has been a shift in the way truffles are consumed, from being eaten whole or in large numbers to being used sparingly to add an element of flavour. Let's backtrack for a moment to see what happened. The Ancient Romans probably ate desert truffles with a milder flavour that was likely more amenable to whole consumption (or perhaps having more potent bioactivity when consumed whole?). There are several recipes involving whole consumption in the 1st Century Roman cookbook *Apicius*.[7] The section on truffles first instructs one to peel the truffles, simmer them in water, then put them on skewers, sprinkle them with salt, and grill them lightly. After pricking them with a fork, one then lets them stand for a few minutes before serving them in one of several sauces:

---

1    Allen, K. and Bennett, J. 2021.
2    Rodrigues, A. 2008.
3    Grigson, J. 1975.
4    Mustafa, A., *et al.* 2020; Allen, K. and Bennett, J. 2021.
5    Allen, K. and Bennett, J. 2021
6    Obligatory cooking pun. Sorry.
7    Grigson, J. 1975.

- Olive oil, *carenum*,[1] wine, pepper, and honey, thickened with starch
- Garum,[2] green olive oil, carenum, wine, crushed pepper, and honey, thickened with starch[3]
- Pepper, lovage, coriander, rue, garum, honey, wine, and a bit of oil
- Pepper, mint, rue, honey, oil, and a bit of wine
- Salt, pepper, chopped coriander mixed with wine, a bit of oil[4]

After the European Renaissance we begin to see truffles added to other dishes like those with poultry. At this point, we are mostly talking about the more potently pungent *T. melanosporum* and *T. magnatum* and their allies, and not *Terfezia* spp. By 1834, Joseph Roques catalogued a variety of recipes that appear to have been firmly established well before he published his work.[5] The recipes do sometimes incorporate the truffles with other foods, but still often involve using up whole numbers of them in various ways. For example:

- **Truffle stew:** Truffles are marinated in oil, sliced, sauteed with oil, butter, salt, pepper, and wine, then served with a bit of lemon juice or bound in egg yolks.
- *Truffes au naturel* : Salt and pepper are sprinkled on the truffles, they are wrapped in paper, topped with bacon, baked, and served warm in a napkin.
- **Champagne wine truffles:** Washed truffles are simmered while covered for a half hour with wine, bay leaf, salt, and bacon, then served under a napkin. (Or you can just wrap them in buttered paper and steam them).
- **Italian truffles:** Truffles are cut and marinated in a saucepan with oil, salt, pepper, parsley, shallots and minced garlic, then cooked on low heat and served with a bit of lemon juice or wine and degreased Spanish sauce.
- **Easy Piedmont truffles:** Oil-marinated truffles are sliced and layered on a silver tray alternating with oil, salt, pepper, and Parmesan cheese, then baked in an oven. (You can also prepare them in a crisp with bacon and place them in a shortcrust pastry to form a sort of pie, then bake for an hour).
- **Truffle pudding:** Sauté two pounds of sliced truffles in water, poultry glaze, Madeira wine, salt, some mignonette, and nutmeg. Line a small buttered bowl with a shortcrust pastry, then add the truffles and seasoning. Cover that with pastry, seal the edges, wrap the bowl with a towel and fix that with a string, and place it in boiling water for an hour and a half. (A dish is also mentioned involving bacon, olive oil, white wine, and nutmeg, served with a bit of lemon juice. One can also use milk to coagulate the truffles and in this manner incorporate them in cheese).
- **Truffle ratafia:** Two pounds of chopped truffles are cold-macerated in *eau-de-vie* (double-distilled fruit brandy) for twenty days. Add Mexican vanilla, sweeten with two pounds of sugar in one pound of water, filter, and keep it in a tightly-sealed flask. It is supposed to be a tonic to cheer saddened spirits and 'awaken languid organs.'

The authoritative *Larousse Gastromique* (1938) lists many standard ways for preparing truffles that echo or replicate the earlier ones. However, it offers ways to use them sliced or in other ways.[6] It particularly suggests that white Italian truffles be sliced and used as a garnish. French author Sidonie-Gabrielle Colette (1873-1954, known simply as 'Colette') had plenty of opinions about how to eat black Périgord truffles. She railed against slicing or incorporating them into *foie gras*, sauces, 'over-larded' poultry dishes, or vegetable dishes, insisting that they be consumed whole:

> *"Away with all this slicing, this dicing, this grating, this peeling of truffles! Can they not love it for itself. If you do love it, then pay its ransom royally—or keep away from it altogether."*

Colette reccomended that one have them on their own, served hot and in 'munificent' quantities. She allowed that one could bathe and cook them in fine wine (not champagne, though), salted or with a bit of pepper. She also wants you to appreciate them for the 'desolate kingdom that they rule' (pebbles and impoverished soils, killing off dog rose and supposedly draining the life from oak trees).[7] Oh, Colette!

A quick perusal of the many online recipes involving truffles will quickly reveal that Colette's advice has been largely ignored. Modern recipes commonly involve shaving or grating sparingly onto warm, neutral flavoured dishes with pasta, rice, cheese or the like and perhaps other vegetables or meat, allowing the warmth to liberate the aroma as one

---

1     Suggested recipe in Part 1.
2     Described in more detail in Part 1.
3     The truffles can be wrapped in caul fat (organ membrane) before skewering, if one prefers.
4     Add some leek to the water when simmering the truffles.
5     Roques, J. 1832.
6     Montagné, P. 1938.
7     Grigson, J. 1975.

appreciates the meal. So what happened? The following is just speculation by yours truly, but one could probably list several potential factors:

- Truffles are dear. As this is being written, black Périgord truffles are in season and their price is hovering around US$ 60- US$85 per ounce (silver is going for around US$24 per ounce). Conspicuous consumption of whole truffles is simply beyond the reach of many consumers.
- On the other hand, adding some shavings or a bit of carpaccio or fragrant oil or infused cheese into a dish is a very practical way to enjoy the flavour. Truffle cultivation has increased supply relative to the early 19th Century, so while it is true that truffles command a very high price, it's also true that an affluent middle class does have opportunities to sample them. Commercial products like carpaccios, flavoured oils and infused cheese are priced within the reach of ordinary consumers and that helps to meet demand. For those selling truffles, it also means increased profits (by effectively stretching supply at prices that increase overall cumulative value).
- Truffles are now known to be fungi. That might seem kind of obvious, but remember that ancient people believed in spontaneous generation of life and considered truffles to therefore be a great source of mystery. People still call them magic, but that's more about enjoying them and using them as an inspiration for culinary creativity.
- Truffles are currently valued for their contribution to flavour, rather than medicinal or aphrodisiac properties. Middle class consumers are more likely to be interested in their culinary aspect and are probably a bit skeptical of their other alleged properties (if they even know or care about them).
- Truffle species are now more clearly delineated. Again, that might seem kind of obvious, but chefs and cooks can better sort out the different ways they want to match differently flavoured truffle species with other ingredients. Copying the way the Ancient Romans treated *Terfezia* spp. simply isn't necessary or practical at this point if one is using *Tuber* spp. Given the extreme pungency of some truffles (the odour of *T. magnatum* can instantly permeate an entire building when unsealed from a container), one may only need a little of the more odoriferous species to appreciate their flavour. In fact, savvy truffle hunters store their truffles along with eggs —the scent is powerful enough to permeate through the porous shells and allow them to cook and enjoy truffle-flavoured eggs even after they've sold off all of their truffles.[1]

**Nutritional content:** As with other truffles mentioned in this series, species mentioned in this section hold more interest for flavour rather than nutritional value—therefore information on nutrients and trace element accumulation is sparse. Truffles such as *T. borchii* are generally found to be a good source of mineral nutrients.[2] A study comparing the biochemistry of *T. melanosporum*, *T. borchii* and other truffles found that protein content in *T. melanosporum* (per 100 g dry weight) declined from about 24g to 21g after 30 days of storage, while proteins in *T. borchii* declined from about 13g to under 6g. In frozen truffles that were then thawed, proteins of *T. melanosporum* declined from about 12.5g to 7.5g after 30 days, while those in *T. borchii* declined from a bit over 9g to a bit over 4g in the same period. For the same 100g dry weight both species had about 2g of carbohydrates.[3]

**Aroma & flavour compounds:** There are dozens and dozens of aromatic chemicals detected in various truffle studies, of which a handful could be considered key, taking both abundance and sensitivity (potency) into account. Regarding infused oils, there are opinions both pro and con, claiming results ranging from acceptable to pungent, off-putting, or obviously artificial or synthetic. Note that some oils are synthetic recreations. As such, one should not expect them to perform as nicely as those derived naturally. Truffle oils used to test the recipes in this series were purported to be from natural sources, and had no obvious artificial or synthetic aroma or flavour.

Descriptions of the flavour of winter truffles in this section and some of their key chemical components are given in the table that follows.

---

1   Luard, E. 2004.
2   Shah, N., *et al.* 2022.
3   Saltarelli, R., *et al.*, 2008.

| Species | Aroma/flavour descriptions | Key aromatic constituents |
|---|---|---|
| *T. borchii* | Pungent, pleasant, garlicky, rich, earthy, nutty | 1-octen-3-ol, alcohols, aldehydes, 2- and 3-methylthiophenes[1] |
| *T. brumale* | Persistent, strong, pleasant, musky, like *T. melansporum*, fermented fruit, bitter yeast, piquant, slightly nutty, woody, bark-like, humus, petroleum | 1,2,4-trimethoxybenzene, 2-methyl-1-butanol, n-heptanal[2] |
| *T. indicum* | Faint, mild, pungent, tuber or truffle-like | 3-methylbutanol[3] |
| *T. melanosporum* | Subtle, hard to define, intense aroma but mild flavour, persistent, pungent, musky green plant, woody, fruity, earthy, chocolatey, cocoa, slightly woody, nutty, sweet, milky[5], slightly mushroomy, slightly peppery & bitter | 3-ethyl-5-methylphenol, 2,3-butanedione, isoamyl alcohol, ethyl butyrate, 2- & 3-methylbutanol, 2- & 3-methylbutanal, dimethyl disulfide (DMDS)[4] |

A 2005 study found that yeast fungi isolated from *T. melanosporum* produce a variety of volatile compounds, including three found in the preceding table (2- & 3-methylbutanol and DMDS). Volatiles produced depended on the yeast species involved (*Candida saitoana*, *Debaryomyces hansenii*, *Cryptococcus* sp., *Rhodotorula mucilaginosa*, and *Cutaneotrichosporon moniliiforme*).[6]

Although freeze-dried truffle products are often offered, a study comparing fresh and freeze-dried *T. melanosporum*, *T. brumale*, *T. magnatum*, and *T. aestivum* has shown that there are significant changes in the profile of volatile compounds present in the fruiting bodies.[7] Another study found significant changes in frozen specimens of *T. melanosporum* and suggested that methional and some phenolic compounds could serve as markers of freezing time. Also, our familiar friend 1-octen-3-one could be a general indicator of freezing.[8] In *T. borchii*, (E)-2-pentenal is released during cold storage, while bis-3-methylbutyl disulphide, 2-methyl-4,5-dihydrothiophene and 3-methyl-4,5-dihydrothiophene are emitted when the species is stored at room temperature.[9]

**Bioactivity:** Bioactive properties reported for *T. melanosporum* and *T. indicum* are shown in Table 3 (p. 6).[10] Among other interesting findings, a study that incorporated *T. melanosporum* into the diets of laboratory mice found that the mushroom could increase physical stamina in swimming trials[11] (similar to the findings for matsutake mentioned in Part 3). A study of antioxidants in *T. melanosporum* found that they decline with storage, but that vacuum-packing or packing in $CO_2$ or nitrogen could extend shelf life and reduce degradation.[12] Researchers taking a look at antioxidants in *T. indicum* found that levels varied depending on the geographic source of the truffle.[13]

---

1    Bellesia, F., *et al.* 2001.
2    Kiss, M., *et al.* 2011.
3    Bellesia, F., *et al.* 2002.
4    Bellesia, F., *et al.* 1998; Culleré, L., *et al.* 2010.
5    According to yours truly.
6    Buzzini, P., *et al.* 2005.
7    Šiškovič, N., *et al.*, 2021.
8    Culleré, L., *et al.*, 2013.
9    Bellesia, F., *et al.* 2001.
10    Tejedor-Calvo, E., *et al.* 2020, 2021; Savini, S., *et al.* 2017; Wu, Z., *et al.* 2021; Li, J., *et al.* 2019.
11    Jiang, X., *et al.* 2018.
12    Savini, S., *et al.* 2017.
13    Li, J., *et al.* 2019.

**Imposters:** Even if you travel a lot, you will be lucky to have someone show you where to find the truffles covered in this chapter (except maybe on some kind of specialized guided tour). There are lots of hypogeous fungi that might resemble the truffle species in this chapter. But since most readers of this book are unlikely to be hunting for these truffle species on their own, it makes more sense to talk about consumer awareness, and the deliberate substitution of inferior truffles (and worse things) for preferred varieties. This is one area where the old maxim *caveat emptor* (let the buyer beware) applies. Supplies of *T. melanosporum* started to diminish during the 20th Century. During the latter part of the century, unscrupulous suppliers took advantage of lax regulation and started to import *T. indicum* from China and present them as the more expensive (but similar in appearance) *T. melansporum*. Unfortunately, the aromas and tastes of these truffles aren't comparable. If you got mild, bland-tasting truffles that lacked the 'punch' that *T. melanosporum* should have, chances were that you were conned. Supplies of *T. indicum* in China have diminished in conjunction with increasing forest disturbance and deforestation in the region, and substitution using that species has recently become less frequent.[1] That being said, the high prices that truffles command continue to be a big temptation for unscrupulous scammers. Sometimes the implications might be serious. A person selling mushrooms on the Facebook Marketplace website recently advertised potentially poisonous *Scleroderma* sp. (earthballs) as truffles. While it is true that earthballs can be confused with truffles by those who are inexperienced, this person blocked messages from knowledgeable people trying to inform him that they were not truffles—he clearly had no interest in being corrected. If you happen to come across some prices for *T. melanosporum* that are too good to be true, they probably are. If you see truffles lacking a species designation, it would probably be a good idea to put your credit card away and check out more reputable suppliers.

One should be aware that some 'truffle' oils and pastes actually use chemicals to artificially replicate products with actual truffles (or products infused with them). Knowledgeable chefs generally pan these products for their artificial, synthetic aroma and flavour.

---

1    Chen, J., *et al.* 2016.

# Recipes
## Truffled Canadian cheddar soup

This soup is just the thing to prepare when you need to fortify yourself for that upcoming hockey game, whether it will be at the skating rink or watching in your living room. Cheddar cheese soup is another meal item associated with Canada,[1] and truffles take it to the next level. This makes quite a lot of soup, so use a big pot. Don't forget to add the warm beer because it lends just the right finishing touch. It can be alcohol-free, but it should be a Canadian brand for true authenticity (take that, wine purists!). Observant readers will notice that this is a pretty cheap and easy way to slip a truffle recipe into a cookbook. Just don't tell that to a Canadian—she will probably blink at you, give you a blank look, and (very politely) tell you that she doesn't really understand your point. She'll then ask, "Have *seen* the price of truffles lately?" Yours truly worried that the truffles might not really stand out or play well with the other ingredients. Happily, this is absolutely not the case. The contribution from cheese infused with Black Périgord truffles is subtle but clear, the perfect thing needed to compliment and complete the other flavours. As an added bonus, truffles are now cultivated in small quantities in B.C.—hopefully they could one day become more commonly available in Canada.[2] So set aside a Molson from that two-four and run off to the charcuterie to get some maple-glazed, smoked Canadian bacon and truffle-infused cheese. And don't forget your toque, eh? It's skookum cold out there.

| | |
|---|---|
| ½ lb | **Canadian bacon,[3] smoked, chopped** |
| 4 tbsp | **Butter** |
| 3 stalks | **Celery, sliced** |
| ½-1 | **Onion, diced** |
| 1 | **Carrot, grated** |
| 2-4 | **Potatoes, medium, diced** |
| 1 cup | **Flour** |
| 3¼ cups | **Chicken stock** |
| 4 cups | **Milk, whole** |
| 2 cups | **Cheddar cheese, truffled (*Tuber melanosporum*), grated** |
| To taste | **Salt[4] & pepper** |
| ⅛ tsp | **Cayenne** |
| ½ tsp | **Paprika** |
| ½ tsp | **Mustard, Dijon** |
| 1 tbsp | **Worcestershire sauce** |
| 1 tbsp | **Mushroom ketchup[5]** |
| ½-¾ cup | **Warm beer, Canadian** |
| 1 | **Green onion, chopped** |
| Some | **Bacon, smoked, chopped** |
| Some | **Paprika (garnish)** |

Cook the bacon in large soup pot (med. heat) until the fat melts (set aside a small bit of it to crisp in a skillet for garnish). Add the butter and melt it. Add diced celery and onion and sauté until onions are translucent. Sprinkle flour over top and mix to make a roux. Add chicken stock and bring everything to a simmer and stir as it thickens, until no longer lumpy. Add milk, grated carrot, and chopped potatoes. Raise the heat a bit and stir. When it reaches a creamy, sauce-like consistency, add the cheese in small increments, carefully watching the heat and not rushing things so that the cheese melts when added. Add the remaining seasonings and the warm beer, and simmer and stir a few more minutes. Serve warm with a sprinkle of paprika, green onions, and chopped bacon as garnish. Serves 3-4.

---

1    Piasecki, B. 2008; Dairy Farmers of Canada. 2021; Taste of Home. 2010.
2    Truffle Association of British Columbia. 2022.
3    Back bacon, not the thin fatty strips.
4    Bacon and cheddar are both fairly salty, so watch the salt. Lots of pepper works, though.
5    Suggested recipe in Part 1.

# Grilled truffled cheese sandwiches

I'll probably get a lot of flak for this one. In my defense, the recipe is based on one that appears in a very fine book covering the use of truffles in *haute cuisine*.[1] It seems that the culinary elite appreciate a good grilled cheese sandwich as much as the next person. At any rate, this is also an easy, delicious way to sample the relatively unadorned taste of truffles if one isn't familiar with it. Truffle-infused cheese is sometimes available in North America, so this might very well be a good way for beginners to experience the flavour of this mushroom (and maybe show off a little).

| Some | 1:1 Olive oil/Melted butter |
|------|------------------------------|
| Some | White bread, firm, crust trimmed |
| Some | Truffle-infused cheddar cheese, (*T. Melanosporum*), sliced |

Preheat oven to 400°F. Amounts will depend on the number of sandwiches (two halved sandwiches take ca. 1 tbsp each of olive oil and butter, 4 slices bread, and maybe 20% of a small brick of cheese). Cut bread slices into halves or quarters or preferred size and shape. Lay slices out on a baking tray and brush tops of half with butter/oil mix. Turn over, lay cheese slices on them, and cover with other bread slices. Brush tops with butter/oil mix and bake 3 min. (until bread toasts light gold and the cheese starts melting). Flip and bake another 3 min.[2] until other side is also a light golden colour. Plate elegantly and serve warm with tomato soup while the cheese is still runny.

---

1    Safina, R. and Sutton, J. 2003.
2    Nor more, no less. Three shall be the number that thou shalt count, and the number of the counting shall be three. Four shalt thou not count, neither count thou two (excepting that thou then proceed to three). Five is right out. See Chapman, G., *et al.* 1975.

# Truffled egg & cheese vol-au-vents

This dish started life as a cheese tart recipe,[1] but there's just something more elegant and tasty about a puff pastry. Cheese, cream, and eggs are all ingredients able to absorb the essence of truffles, so this is a fine way to focus on what they can do for flavour in a tasty *hors-d'œuvre*. Have fun building your next-level pastry treat!

| | |
|---|---|
| 4 | Eggs |
| ½ cup | Table cream |
| 2 tsp | Truffle-infused oil (*T. melanosporum*) |
| ¼ tsp | Salt |
| ⅛ tsp | White pepper |
| ¼ tsp | Seasoned mushroom powder[2] |
| 1 cup | Emmental cheese, shredded |
| 12[3] | *Vol-au-vent* puff pastries, 8-cm diam., frozen |

Partly defrost and bake the puff pastries on a tray according to package instructions (perhaps 18-20 min. at 400°F). Use a whisk to beat and blend the eggs in a bowl, followed by the cream, truffle oil, salt, and white pepper. Place the pastries on a wire rack over a tray or counter so that the bottoms won't become soggy if you spill some filling. Fill the vol-au-vents loosely with the shredded cheese and very carefully pour the egg and cream mix into the pastries and over the cheese. Replace the pastry lids and bake the pastries at 350°F, long enough to melt the cheese and set the eggs (ca. 7 min.). Serve warm.

---

1    Safina, R. and Sutton, J. 2003.
2    Suggested recipe in Part 1.
3    They often come 6 per package.

# Imperial stuffed eggs

This is a modern recipe, but the history of devilled eggs harks back to the Roman Empire, where hard-boiled eggs were chopped up and covered with a sauce having various ingredients. A recipe for stuffed eggs appeared in 13th Century Andalusia (Islamic-ruled Iberia). Usually, the halves of the eggs were held together after being stuffed. Eventually, recipes developed that involved frying them in butter after stuffing. Seasoning the stuffing with Cayenne pepper added some heat that caused the food to be 'devilled.' Modern recipes finally arrived with tamer stuffing seasoned with paprika, along with special trays having indentations to hold the eggs.[1] Here we will forgo *Capsicum* spices and use mushrooms (surprise!) in the form of duxelles with thyme, and mix that with the yolks and mayonnaise (or sour cream and dill—either way is savory and delicious). Technically then, these eggs are no longer devilled. But we'll also be harking back to Imperial Rome and adding some essence of those fabled treats arising from Jove's thunderbolts: truffles.

| | |
|---|---|
| 6 | Eggs |
| ½ cup | Duxelles[2] (*Agaricus bisporus*), (w/ thyme) |
| To taste | Salt & pepper |
| ½ tsp | Olive oil infused w/ truffles (*Tuber melanosporum*) |
| 1-2 tbsp | Sour cream or mayonnaise |
| 2 tsp | Dill, chopped (if using sour cream) |

Put eggs in a large pot, cover with 1" water, and cook in simmering water for about 12 min. (until they are hard-boiled). Chill the eggs in very cold or ice-cold water 30 min. Heat the duxelles if they are frozen. Carefully peel the eggs (perhaps with the help of a small spoon sliding under the shell), and slice them into halves. Remove the yolks to a bowl and set the whites aside in a tray. Mix the duxelles with the yolks and add the salt and pepper, using a fork to mash everything together. Mix in the sour cream (if using dill) or mayonnaise (if using thyme) bit by bit (use just enough to make a nice, smooth blend that holds together well). Cool the filling and stuff the egg whites carefully with a small spoon. Serve chilled as soon as possible, do not store. If you are bringing these to an outdoor gathering, avoid setting in direct sunlight or in warm areas that might quickly cause them to spoil.

---

1    Miller, M. 2023.
2    Suggested recipe in Part 1.

# Pittsburg Potatoes

Somewhere in the territory between Potatoes Romanoff [1] and Scalloped Potatoes *au gratin* you will find a hearty dish from the mid-western U.S. called Pittsburg Potatoes. Its origins are a little mysterious, but can perhaps be pinned down with a little help from the letter 'h.' About 1915, a Presbyterian Ladies Aid Society cookbook from Gary, Indiana published a version of this dish called 'Pittsburg Potatoes.'[2] Pittsburgh, Pennsylvania might leap to mind, but that city had its 'h' as of 1758, made official in the city charter of 1816. In 1891 the U.S. Board of Geographic names tried to take away the 'h' to standardize the spelling of various 'burgs' in the U.S. They erroneously believed that the charter of the Pennsylvanian city lacked the 'h', based on copies of the document that mistakenly omitted it. After protests, the 'h' was reinstated in 1911. So where was 'Pittsburg' mentioned in the 1915 publication then? About 75 miles SSW of Gary, Indiana there is an unincorporated town called Pittsburg in Tippecanoe Township of Carroll County. Likely named after Pittsburgh, Pa., it was founded in 1838 and was once a thriving mill town and centre of commerce along the side channel of a busy canal. There were grain warehouses, flour mills, oil and saw mills, wool carding and fulling operations, foundries, machine shops, taverns, and various shops for coopers, cabinet and furniture makers, blacksmiths and shoemakers. It declined when the Wabash Railroad came in 1856 and further in 1881 when farmers dynamited a nearby dam and destroyed the canal. Pittsburg officially lost its 'h' in 1894 and never regained it. Given the mid-western focus of the dish, yours truly believes that the recipe name refers to Pittsburg, Indiana. Pittsburg Potatoes didn't originally include mushrooms, but there are some current recipes that use them.[3] Other tasty upgrades added here include truffle-infused cheddar cheese and chopped bacon (making this dish even tastier and nicer to have on a cold winter day).[4]

| | |
|---|---|
| 4 cups | Potatoes, diced |
| 1 | Shallot or onion, chopped |
| 4 cups | Water |
| ¼-½ cup | Pimentos, drained |
| 4 strips | Bacon |
| 3-4 tbsp + | Butter |
| 5 tbsp | Flour |
| 1 cup | Milk |
| To taste | Salt & pepper |
| To taste | Paprika |
| 1 cup | Cheddar cheese, grated |
| 1 cup | Truffle-infused cheese (*T. melanosporum*), grated |
| 1 can | Mushroom soup (*Agaricus bisporus*)[5] |
| 1½ tbsp | Bread crumbs |
| 1½ tbsp | Bacon, fried, chopped |

Preheat oven to 375°F. Put diced potatoes and chopped shallot or onion in a pot with salted water and boil 5-6 min. Add pimentos and boil 5-6 min. until almost tender, then drain (conserve the starchy water). Put potatoes and pimentos in a buttered baking dish and set aside. Sauté the bacon strips in a large skillet until crisp, remove, chop and set aside (keep the melted fat in the skillet). Make a roux in a the skillet (med. heat) by vigorously mixing butter, flour, milk, ½ cup of the conserved starchy water, salt, pepper, and paprika. Add half of the cheese and stir until melted. Add the rest of the cheese and some conserved water as necessary until achieving a thick sauce that nicely coats and sticks to a spoon. Mix in the mushroom soup and chopped bacon (add a bit more of the conserved starchy water if this becomes too thick). Mix the cheese and mushroom sauce in with the potatoes, shallot, and pimentos. Sprinkle the top with bread crumbs and/or crispy bacon bits, dot with butter, and bake until done (about 20 min.). Cool a bit before serving.

---

1    Suggested recipe in Part 3.
2    Ladies Aid Society, ca. 1915, cited by Powell, G. 2022.
3    WesO8647. 2011; Beeler, J. 2000.
4    There is a similar cheesy, *au gratin* dish called 'Funeral Potatoes' found further west in areas settled by Mormons. The dish was popularized by another woman's organization, the Mormon Relief Society. The duties of a Mormon include attending to the needs of the bereaved (e.g. meals), and that dish makes convenient use of items that should be in any Mormon's 3-month food supply (a requirement that appeared after the era of the Great Depression).
5    Many recipes also include sour cream, however there is plenty of creaminess in the soup.

# Tomato-less pasta

Yes, we're actually talking about roast pepper pasta sauce. One of my family members doesn't like pasta made with tomato sauce, so I turned to roasted peppers for help. You can roast them in your oven, of course, but jars of roasted red peppers are conveniently available in many B.C. grocery stores. They can be used to make a tasty red sauce for a Mediterranean-style pasta dish,[1] without anyone ever having to go *near* a tomato. Popular recipes for this sauce emphasize vegetarian ingredients and lower fats (less cooking oil, less cheese, and milk rather than cream) blended into a totally smooth sauce. The vegetarian aspect remains in this recipe, but there's bit more richness (we all like cheese) and added truffle and mushroom goodness (for flavour and texture). Parmesan is favoured where cheese is present in other recipes, but we're using pecorino here. Parmesan and pecorino are both aged, hard, salty cheeses (Parmesan is made with cow's milk, while pecorino comes from sheep's milk). Not to worry, pecorino cheese gives this dish lots of zip. It's another excellent alternative!

| | |
|---|---|
| 1 jar | Roasted red peppers |
| ½ cup | Table cream |
| 2 | Shallots |
| 3 cloves | Garlic |
| ½ tsp | Oregano |
| ½ tsp | Smoked paprika |
| ½ tsp | Basil |
| ¼ tsp | Italian seasoning |
| 2-3 tsp | Truffle-infused oil (*T. melanosporum*) |
| To taste | Salt & pepper |
| 6 cups | Water, lightly salted |
| 250 g | Farfalle rigate (bowtie pasta) |
| 2 tbsp | Sunflower oil |
| 1 cup | Creminis (*A. bisporus*), sliced |
| ¼ cup | Pecorino Romano or Parmesan cheese |
| 1 cup | Spinach leaves, chopped or torn |

Puree drained peppers, cream, garlic, shallots, oregano, smoked paprika, truffle oil, salt, and pepper in a food processor or blender. Set aside for a moment. Bring water to a boil in a large pot, add pasta, and cook according to package instructions (ca. 8-9 min.). Meanwhile, heat oil in a skillet (med. heat) and sauté shallots and garlic until shallots are soft and transparent and garlic aromatic (2-3 min.). Add the mushrooms and cook until softened and browned (ca. 4-5 min.). Stir in pepper purée and simmer 5 min. Add enough mushroom broth[2] to reach the desired consistency (perhaps not the entire cup). Remove from heat and quickly stir in cheese so that it melts. Adjust seasonings. Drain pasta and briefly rinse with a bit of cold water to keep it *al dente*. Mix pasta and sauce so that the pasta is coated, top with spinach and serve warm.

---

1    Drummond, R. 2013; Allen, L. 2020; Jawad, Y. 2022; Karadsheh, S. 2023.
2    Alternatively, reserve salted water after draining pasta and use that.

# Pasta frittata

Breakfast time is over and it's not quite lunchtime yet, so you're hungry for a nice brunch or early lunch. You have a bit of time and a stock of truffle oil to add some elegance, so why not make a quick pasta fritatta? If you don't have an oven-ready skillet, don't worry. You can still manage with a regular skillet and a baking dish (pictured) with a bit more time and finesse. This frittata has all those nice, tasty breakfast things (eggs, bacon, and cheese) combined with enough lunchtime carbohydrates (in the pasta) to give you energy throughout the afternoon.[1] And did I mention it's quick?

| | |
|---|---|
| 6 cups | Water |
| 200 g | Spaghettini noodles |
| 3 tbsp | Butter |
| 4 strips | Bacon |
| ½ tsp | Truffled olive oil (*T. melanosporum*) |
| 4 | Eggs, beaten |
| 1 | Onion, diced |
| ¼ cup | Parmesan cheese, grated |
| ½ cup | Gruyere cheese, shredded |
| 1 tsp | Smoked paprika |
| 1 pinch | Cayenne |
| To taste | Salt & white pepper |

Preheat oven to 350°F. Bring lightly salted water to boil in a pot, break pasta in half,[2] and cook pasta until it is softened but not yet *al dente*, perhaps ca. 5-6 min. (you will be baking it longer in the oven). While the pasta is boiling, melt 2 tbsp butter in a skillet. Fry the bacon (med. heat) 2-3 min. per side (until it reaches the preferred level of crunchiness without burning). Remove the bacon and place between paper towels to blot away excess grease, and then take away the towels and chop the bacon. Add the onion to the skillet and sauté until the onion is limp and translucent. Remove from the skillet and mix the bacon and onion and cheese with the eggs. Drain the pasta, then toss with the remaining 1 tbsp butter and the truffle-infused oil in a large bowl. Add the egg, onion, and bacon mix and season with salt and white pepper. Pour the mix into the a shallow baking dish, allowing some pasta ends to stick up (to become crunchy). Sprinkle some more Parmesan cheese on top and bake until the top firms up and lightly browns (ca. 15-18 min.). Remove from oven and serve slices warm.

---

1     Bittman, M. 2018.
2     Don't worry about violating the usual proscription against breaking pasta. You aren't trying to fit pasta into a pot for a regular pasta dish, you are doing this to improve its texture in the fritatta.

# Deluxe penne & cheese

Canadians are crazy about their KD (Kraft Macaroni and Cheese Dinner ®). Yours truly certainly views all versions of tubular pasta and cheese as a major essential food group. Derived rom a published recipe, here is an elegant dish[1] that will hopefully inspire you to take KD to the next level. It substitutes penne pasta for macaroni noodles and truffle-infused cheddar cheese for dried cheese powder. It also includes another important Canadian favourite: bacon.

| Basic dish | |
|---|---|
| 6 cups | Water |
| 225 g | Penne pasta |
| 6 slices | Bacon |
| 3 tbsp | Olive oil |
| 1 clove | Garlic, minced |
| 2 tbsp | Flour, all-purpose |
| 1 cup | Milk |
| 1-2 cups | Cheddar cheese, regular, shredded |
| ½ cup | White cheddar cheese, Truffle-infused (*Tuber melanosporum*) |
| To taste | Salt & pepper |

**If baking with light crust:**

| | |
|---|---|
| ¼ cup | Parmesan cheese |

**If baking with fluffy topping**

| | |
|---|---|
| ¾ cup | Parmesan cheese |
| ¼ cup | Breadcrumbs |

If you will be baking for a light crust, preheat oven to 350°F (see below). Add about lightly salted water to a large pot. Bring the water to a boil and cook the penne according to package instructions (perhaps 9-10 min.). Meanwhile, sauté the bacon in a skillet (med. heat) until it reaches the preferred amount of crispness. Remove from the skillet and chop (leaving the melted fat behind). Add olive oil to the melted fat and sauté the minced garlic (ca. 1 min.), until aromatic. Stir in the flour to form a smooth and (after 1-2 min) bubbling roux. Gradually whisk in the milk over 2-3 min. until you have a relatively thick white sauce. Lower the heat and add the cheeses and bacon, stirring until the cheese melts into the smooth sauce (2-3 min.). Season with salt and pepper. Drain the pasta, rinse it briefly in a bit of cold water to keep it at the *al dente* stage, and return it to the pot. Add the sauce to the noodles, mixing until it is all coated. At this point it will taste great and can be served warm with a salad (see photo). But if you prefer to have it baked with a crust[2], put everything in a baking or casserole dish, sprinkle top with Parmesan cheese and bake for 25-30 min. (until a bubbling, lightly browned crust forms at top). For a fluffy, slightly moist topping, top instead with a thicker 3:1 mix of Parmesan cheese and breadcrumbs before baking. Plate while it is warm (serves 4-5).

---

1    Safina, R. and Sutton, J. 2003.
2    This is also something to consider if you are reheating leftovers and want to serve it a different way.

# Eggs Benny w/ truffled Hollandaise sauce

Savvy European truffle hunters keep their truffles with a basket of eggs until they can sell them, thus infusing the eggs with the wonderful aroma and providing an extra bonus for the lucky finder. If you have fresh truffles, you can actually use them to infuse eggs in a jar or similar container with a close-fitting lid. You wrap the truffle in a bit of paper towel (for moisture control), put it in a tea-ball or similar containment open to diffusion, put the eggs (or cheese, etc.) in the jar, seal it tight, and refrigerate. Change the paper towel the next day and after another day in the jar the fats in the egg yolks or cheese will have absorbed some truffle aroma (and you can still use the truffle for something else). On the other hand, there may not be any need for any of this if you have truffle-infused oil on hand and you're prepared to make something like eggs Benedict.[1] You can slip the oil into the Hollandaise sauce and enjoy the same kind of treat.

## Hollandaise sauce

| | |
|---|---|
| 3 cups | Water |
| 2 | Egg yolks |
| ¼ cup | Cream |
| 1 tsp | Apple vinegar |
| ½ tsp | Lemon juice |
| ½ tsp | Dijon mustard |
| 4 tbsp | Butter, melted |
| ½ tsp | Truffle-infused olive oil (*T. melanosporum*) |
| To taste | Salt & white pepper |

## Eggs Benedict

| | |
|---|---|
| 4 cups | Water |
| 4 | Eggs |
| 4 strips | Bacon, cut in half |
| 2 | English muffins |

Melt the butter and scoop off the solids to clarify it. Meanwhile, bring 3 cups water to a boil in a double boiler (or if you don't have one, in a large pot). Reduce heat to a simmer. In a heat-resistant bowl, whisk together the egg yolks, cream, vinegar, lemon juice, and mustard until the mix is as smooth and frothy as you can manage. Place the bowl over the simmering water in the double boiler (if you are using a pot, insert a ramekin into the pot so that the bowl can rest on it well above the water). Allow the sauce to heat for about 5 min. while occasionally whisking (do not allow it to come near a simmering temperature—ideally it should remain around 65°C, a temperature at which you can still handle the bowl). Slowly add a thin stream of the melted clarified butter and truffle-infused olive oil as you whisk and let the sauce thicken (it may take ca. 10-15 min. to properly thicken—when a little is whisked in a shallow bowl you should occasionally see the bottom). Keep whisking frequently to ensure a smooth sauce. Remove the Hollandaise sauce from heat and season with salt and white pepper. Meanwhile, bring 4 cups of water to a boil in another pot or large sauce pan and reduce heat to a simmer. Also, sear the bacon in a skillet 2-3 min. per side (med. heat). Add the eggs to the boiling water and poach them (ca. 2-3 min., until whites have firmed up a bit). Use poaching pods or containers if you wish. Also, slice and toast the English muffins. Plate the muffins halves and top them with bacon, followed by poached egg (remove with a slotted spoon), followed by the Hollandaise sauce. Serve while warm. Serves 2-4.

---

1    Lydon, K. 2014; TRUFF. 2024

# Truffled bacon & chicken Alfredo

During the early to mid 1900s, Alfredo Di Lelio featured the original version of this pasta dish at his restaurant in Rome. Fresh fettuccine was coated with melted butter and then Parmesan cheese added, resulting in a rich, melted cheese sauce. Diverse recipes for this dish have evolved since them, and this is one of them. Part 2 of this series has a recipe for quick chicken Alfredo using commercial Alfredo sauce (in the *Agaricus* section). We'll try not to cut corners here, which means making the Alfredo sauce from scratch. We'll also enhance things a bit by including truffle oil, bacon, and mushroom slices. If you've been using commercial sauce up to now, this is a chance to indulge your curiosity and see how things turn out if you make it yourself. Don't worry. It's really not that hard and your dish will turn out to be elegant and delicious.

| | |
|---|---|
| 6 cups | Water |
| 250 g | Fettuccine or linguine pasta |
| 1 tbsp | Butter |
| 3 tbsp | Olive oil |
| 2 tsp | Truffled olive oil (*T. melanosporum*) |
| 3 strips | Bacon |
| 2 cloves | Garlic, minced |
| 1 cup | Creminis (*A. bisporus*), sliced |
| 1 cup | Table cream |
| 1½ cups | Parmesan cheese, shredded |
| To taste | Salt |
| 4 | Chicken breasts, small, bones & skin removed, chopped |
| 2 | Green onions, sliced |

Add the water to a large pot, along with a pinch of salt and a dash of olive oil. Bring the water to a boil and cook the pasta according to package instructions (perhaps 8-10 min.). Meanwhile, sauté the bacon in a skillet (med. heat) and remove and chop finely when crispy. Add and cook the cubed chicken until the centers of the pieces are done (white when sliced with a knife, ca. 5 min. with turning). Remove the chicken and set aside in a bowl with the chopped bacon. Add the remaining olive oil in the skillet and sauté the garlic until it is aromatic (ca. 1-2 min.). Add the mushrooms and continue to cook until limp and browned (ca. 5 min.). Reduce heat and stir in the cream and allow to simmer 2-3 min. while the liquids reduce. Stir in the cheese and allow the sauce to simmer and thicken another 2-3 min. Adjust salt seasoning. Stir in the cooked chicken, bacon and green onion. Remove from heat when the chicken is warmed through. Drain the pasta and briefly rinse with a small amount of cold water to keep the pasta at the *al dente* stage. Stir the remaining butter into the pasta. You can mix the sauce with the pasta while it's in the pot and then plate it that way, or plate a bed of pasta first, topped with the sauce. At any rate, serve warm and offer with a vegetable and green garden salad. Serves 3-4.

# Truffled chicken roulade

You might come across gourmet recipes for chicken that refer to roulade, ballotine, and galantine (sometimes conflating these terms). A roulade is boneless meat wrapped around a filling made with vegetables, meat, cheese or dairy products, mushrooms, etc. Its size can vary. Ballotines and galantines are both larger types of roulade, made with an entire de-boned bird, fish, etc. Galantines are traditionally stuffed with forcemeat, poached in a gelatinous stock, and served cold with an aspic. Ballotines are like that, but served hot. Roulade recipes can become very fancy and artistic, augmented with various decorative sauces and elements. This recipe, on the other hand, takes a basic approach with a smaller portion of meat. It combines several roulade recipes for chicken and fillings having truffles and mushrooms.[1] We'll cut the bones from chicken breasts here, because it's cheaper to purchase breasts with the bone in, and we can use the extra trimmings in the filling. Take care—you may be tempted to try ever fancier roulades with decorative elements and multiple sauces of contrasting colour and flavour. Roulades can be a 'gateway drug' to more elegant gourmet cooking!

| | |
|---|---|
| 3 tbsp | Butter |
| 2 tsp | Olive oil |
| 1 | Onion, diced |
| 4 cloves | Garlic, minced |
| 3 cups | Creminis, sliced (*A. bisporus*) |
| 1 cup | Porcini, sliced (*B. edulis*) |
| To taste | Salt & pepper |
| 1 | Chicken breast, w/ bone[2] |
| ½ cup | Goat cheese |
| 1 tbsp | Dijon mustard |
| 1 | Egg white |
| 2 tsp | Truffle-infused olive oil (*T. melanosporum*) |
| To taste | Sage |
| To taste | Thyme |
| 10 | Bacon slices |
| 1 cup | Chicken stock |
| 1½ tbsp | Flour |
| ¼ cup | Table cream |
| 8 sprigs | Parsley |

Preheat oven to 375°F. Melt 2 tbsp butter in a skillet (med. heat), add olive oil, and sauté the onion and garlic until onion is soft and translucent and garlic aromatic (2-3 min.). Add mushrooms and sauté 4-5 min. (until browned)and season with salt and pepper. Place half in a bowl and the rest in a blender. Trim away bone from the chicken breasts and remove the skin and extra parts. Trim the ends of the meat to be square. Add the trimmed parts to the blender and pureé this filling, along with goat cheese, mustard, salt and pepper, truffle oil, and egg white. Make a shallow cut down the centre of each chicken breast and cut and open up and out like a butterfly. Pound the breasts so that they are thinner and even. Sprinkle on salt and pepper, sage and thyme. Spread the filling onto the breasts and roll them up. Lay the seam side down on a layer of bacon and fold the bacon around. If you desire, sear the roulades briefly in the skillet (surface only). Place on a wire rack on a baking tray lined with parchment and bake 25 min. (core temperature should reach at least 165°). While baking, puree remaining mushroom mix in blender. Add to skillet and simmer (med. heat). Stir in flour, and after 1-2 min. slowly stir in stock. Simmer until sauce thickens, add cream, and continue cooking and occasionally stirring until sauce is thick. Remove roulades from the oven. Ladle a dollop of sauce in the centre of each plate, and lay slices of roulade on sauce. Blitz parsley in the blender with a bit of cream and/or water. Serves 4-5.

---

1     Four Magazine. 2014; Bruxel, A. 2021; Daniel. 2022; Richa. 2023; Wiringa, J. 2023. Chicken Farmers of Canada. 2024.
2     Many stores sell chicken breasts that are already halved. If so, the two halves count as one breast here. If the breast is whole, split before proceeding.

# Quick truffled tuna pasta

It's quick, it's simple, it's delicious, and it has truffled tuna.[1] What are you waiting for?

| | |
|---|---|
| 6 cups | Water |
| 200 g | Pasta (linguine) |
| 2 tbsp | Olive oil |
| 1 clove | Garlic, minced |
| 1 can | Tuna (in brine) |
| 1 stalk | Celery, diced |
| 3 tbsp | Chicken stock, concentrated |
| ½-1 tsp | Lemon juice |
| 2 tbsp | Parsley, chopped |
| ½ tsp | Truffled olive oil (*T. melanosporum*) |
| To taste | Salt & pepper |
| 1-2 tbsp | Parmesan cheese, grated |

Bring lightly salted water to a boil in a large pot and cook the pasta according to package directions (ca. 9-10 min.). While the pasta is cooking, heat the regular olive oil in a skillet (med. heat), add the garlic, tuna (with juices), celery, lemon juice, and parsley. Cook until the liquids reduce (ca. 3-4 min.). Drain the pasta and briefly give it a little splash of cold water to keep it at the *al dente* stage. Remove the tuna from heat and stir in the truffle-infused oil and Parmesan cheese. Season with salt and pepper. When the cheese has melted, mix the tuna with the pasta. Serve warm in bowls, garnished with a sprinkle of Parmesan cheese on top. Serves 2-3.

---

1    Based on a receipe by Bull, N. 2024.

# Truffled trout Florentine w/ creminis & pasta

A dish prepared in the style of Florence, Italy usually involves a main ingredient accompanied by spinach in a Mornay sauce (béchamel sauce with grated or shredded cheese). You can have chicken Florentine, eggs Florentine, and even salmon Florentine.[1] That's all wonderful, but today we're having steelhead trout, and adding in some creminis and pasta.[2] Steelhead trout (*Oncorhynchus mykiss*) are in a category of their own. Some stay in inland waters while others spend part of their time in the ocean. The flavour of the meat is refined and subtle, without the strong wild fish flavour found in other trout species. The taste is similar to salmon, only more sublime and savory. It's the perfect match for a truffled Mornay sauce in this very delicious and satisfying dish.

| | |
|---|---|
| 6 cups | Water, salted |
| 200 g | Linguine |
| 2 | Steelhead trout fillets (or other trout, or salmon) |
| 3-4 tsp | Truffle-infused olive oil (*Tuber melanosporum*) |
| 3 tbsp | Olive oil |
| 3 cloves | Garlic, minced |
| 1 cup | Creminis (*A. bisporus*), sliced |
| 1 tbsp | Flour |
| ¾ cup | Chicken stock |
| ½ cup | Half & half cream |
| To taste | Salt & pepper |
| ½ cup | Emmenthal cheese, grated[3] |
| 1 cup | Spinach, chopped |
| 1 tbsp | Butter |
| Some | Chives |

Bring salted water to boil in a pot and cook pasta according to package directions (perhaps 8-9 min.). Meanwhile, sprinkle salt and pepper on the fillets along with 1 tsp truffle oil, and place (seasoned side down) in a large skillet. Season the other sides with salt and pepper along with 1 tsp truffle oil, and cook the fish ca. 5 min. per side (turning once). Remove and set aside on a covered plate to keep warm. Drain pasta when cooked, quickly rinse with a small bit of cold water to stop cooking at the *al dente* stage, and return to pot. Stir butter into the pasta and cover to keep warm. Heat the remaining truffle-infused oil along with 2 tbsp olive oil in the skillet and sauté the garlic until fragrant (ca. 1 min.). Add mushrooms and sauté until limp and browned (ca. 4-5 min.). Stir in flour, cook about a minute while stirring, then slowly stir in the stock and cream. Season the sauce with salt and pepper. Lower heat and simmer until liquids reduce and sauce reaches desired thickness. Add cheese and stir until it melts. Add the spinach and cook while stirring until the spinach is limp (ca. 1 min.). Plate the pasta, top with the spinach and Mornay sauce, then lay the fillet on that, garnish with chives, and serve warm with salad and a vegetable (e.g. French-style green beans). Serves 2-3.

---

1  Awada, R. 2024.
2  Or salmon. See Yates, L. 2023.
3  Cheeses used in Mornay sauce include Gruyère, Emmenthal, Parmesan, or white cheddar. The Bavarian-style Emmenthal cheese tested here was superb.

# Creamy truffled pork pasta

This dish is derived from an Italian one called *Pasta alla Norcina*. If you want to prepare that dish the customary way (or find out more about it), you can check out something closer to the authentic thing in the Summer Truffle section of Part 2. The version in Part 2 uses *Tuber aestivum*, but *T. melanosporum* may work even better as an ingredient. With that idea in mind, the dish on this page offers a slightly different way of reaching the same level of delight and satisfaction. Preparation is streamlined by using frozen mushroom slices left over from making broth. Also, the porcini is already in the pasta. Overall, you get three great mushrooms working together: button mushrooms, porcini, and truffles.

| | |
|---|---|
| 1 lb | Button mushrooms (*Agaricus bisporus*), sliced, cooked, frozen |
| 2 cups | Water |
| ½ cup | Dry white wine (Riesling) |
| 3 tbsp | Soy sauce |
| 3 tbsp | Mushroom ketchup[1] |
| 1 tsp | Seasoned mushroom powder[2] |
| 225g | Italian pork sausage[3], crumbled |
| ½ cup | Cream |
| 1 sprig | Chopped parsley |
| 1 cup | Cheddar cheese, grated, truffle-infused (*Tuber melanosporum*) |
| 2 | Green onions, sliced |
| 6 cups | Lightly salted water |
| 150 g | German egg noodles, w/ porcini (*B. Edulis*) flavour |
| Some | Parmesan cheese, grated |

Add mushrooms, 2 cups water, wine, soy sauce, mushroom ketchup, and seasoned mushroom powder to a pot. Bring to a boil, then reduce heat and simmer partly covered 40 min. As liquid reduces, gradually stir in the cream, and chopped parsley. Add the truffle-infused grated cheddar and stir until it is melted. When liquids reduce to the desired thickness, stir in the green onion slices, cover the pot and keep warm. Meanwhile, bring about 6 cups to a boil in another large pot. Boil the noodles according to package instructions (about 9 min.), until pasta is limp but not soggy. Drain the noodles, mix them with the sausage/mushroom sauce, and plate. Sprinkle Parmesan cheese on top, and serve with Caesar or garden salad and garlic bread. Serves 4-5.

---

1    Suggested recipe in Part 1. If you don't have any handy, try 1 tbsp Worcestershire sauce, 1 tbsp tomato ketchup, 1 tsp mustard, and an extra 1 tsp seasoned mushroom powder.
2    Suggested recipe in Part 1.
3    If not available, suggested recipe in Part 1.

# Spanish-Italian pasta

The proximity of Spain and Italy in Europe means that their dishes sometimes blend or overlap. This spicy dish is a fusion of Spanish and Italian (and French) cuisine. The chorizo and Padrón peppers are Spanish ingredients. Meanwhile, the spaghetti noodles are Italian, but they also resemble popular Spanish and Mexican noodles called *fideos*. Adding butter to the noodles is a trick straight from the playbook of Chef Ettore Boiardi (a.k.a. Chef Boyardee), who cooked in the style of northern Italy. In contrast, adding garlic to tomato sauce has more affinity to southern Italy. French ingredients are also present. The Syrah (a.k.a. shiraz) wine used here mostly comes from the Rhône Valley—it has a distinctive, pleasing, slightly sweet flavour that blends well in sauces. Black Périgord truffles also hail from France, although some truffle production has recently shifted to Spain, so they could also be from there. It's tempting to think of Roma tomatoes as an Italian ingredient, but this particular tomato variety was actually developed by the USDA in the 1950s in Beltsville, Maryland. At any rate, this blend is a nice way to perk up your usual spaghetti dinner, and it's also a delicious way to try out that truffle oil and premium Italian pasta in your cupboard. *Buen provecho et buon appetito*![1]

| | |
|---|---|
| 10-12 | Roma tomatoes |
| 2 tbsp | Olive oil |
| 2 | Garlic cloves, minced |
| 1 | Onion, diced |
| 4 | Padrón peppers, diced |
| 200 g | Chorizo[2] |
| 2 | Basil leaves, fresh, chopped |
| 1 tsp | Oregano |
| 1 tsp | Thyme |
| 1 tsp | Italian seasoning |
| To taste | Salt & pepper |
| 3 tbsp | Red wine (syrah) |
| ½ cup | Mushroom stock |
| 2 tsp | Truffle-infused oil (*T. melanosporum*) |
| 1 cup | Parmesan cheese, shredded |
| 100 g | Spaghetti, premium[3] |
| 1 tbsp | Butter |

Preheat oven to 400°F. Cut 2 crossed slits in the tops of the tomatoes, place them on a baking tray, and roast 15-20 min. Meanwhile, heat oil in a large skillet (med. heat) and sauté onions and garlic until the onions are soft and semi-translucent and the garlic aromatic (ca. 2-3 min.). Peel any casing from the chorizo, dice and add it, and continue to cook and stir occasionally until the meat is seared on all sides (2-3 min.). Add the herbs and seasonings, cook another minute or so, then deglaze with the wine and add the stock. Reduce heat to a simmer. Put the tomatoes in a large sauce pan or a stew pot and give them very quick rinse with cold water. The skins should separate and they should cool enough to handle safely. Peel away the skins and any tougher greenish pulp, then mash them. Manually remove any remaining skins or tougher bits of pulp. Add the mix from the skillet and stir in the truffle-infused oil. Continue to cook, letting the sauce simmer 10-15 min. Meanwhile, bring 6 cups of lightly salted water to a boil in a large pot and cook the pasta according to package directions (perhaps 8-9 min. until it is *al dente*). Drain the pasta and rinse very briefly with a dash of cold water to stop the pasta at the al *dente* stage and stir in the butter when you put it back in the pot. Remove the sauce from heat, stir in Parmesan cheese, and either mix with the pasta or plate the pasta and top it with the sauce and a bit more Parmesan cheese. Serve with a garden salad and dressing. Serves 4-5

---

1   By the way, if you're looking for other spaghetti sauces to try, check out the 'Solanaceous sauce' in the *Agaricus* section of Part 2.
2   If you can't find it, there is a recipe for making chorizo on p. 83. That recipe is based on ground pork (Mexican style), for crumbling, not dicing.
3   Also try this sauce with porcini- and truffle-stuffed triangoli—it's a delicious combination! For more about triangoli see the bolete section of Part 3.

# Pork tenderloin (Richard's Way)

There are lots of ways to prepare a mushroom sauce to go with pork tenderloin.[1] Tenderloin is a premium cut of meat, so people can get very fancy with how they dress or prepare it. For example, you can stuff the tenderloin with a filling. That means cutting a slit lengthwise and laying the tenderloin open like a butterfly and then pressing or pounding it flat and even. Then you put a layer of thick sauce on top and carefully roll up the meat so that it encloses the filling. You tie it together with string so that it won't flop open (or hold it together with toothpicks or small skewers), then bake it in the oven at a prescribed temperature for a recommended time. When you take it out of the oven, you check it with a meat thermometer to make sure that the centre is done (recipes never really tell you how long you have to put it back in if it isn't). When it's sliced and served, it's a lovely dish. But some won't fancy anything that elaborate. You might want a less fussy dish that's still savory and delicious. Maybe you have some truffle oil and some king oyster mushrooms. If so, you've come to the right place. Because this is the way we cook pork tenderloin at Chez Winder.

| | |
|---|---|
| 1 tbsp | Truffle-infused oil (*Tuber melanosporum*) |
| ¼ cup | Vinegar, blackberry-flavoured |
| ¼ tsp | Salt |
| ⅛ tsp | Pepper |
| 2 | Garlic cloves, minced |
| 1 | Shallot, chopped |
| 350g | Pork tenderloin[2] |
| 3 tbsp | Butter |
| 2 tbsp | Avocado oil |
| 2 cups | King oyster mushrooms (*P. eryngii*), sliced |
| 3 tbsp | Red wine (syrah/shiraz) |
| ½ cup | Vegetable stock |
| 1 tbsp | Oyster sauce |
| 2 tsp | Soy sauce |
| 1 tbsp | Worcestershire sauce |
| 2 tbsp | Dijon mustard |
| 1 cup | Parsley, chopped |
| ½ cup | Table cream |
| 2 tbsp | Flour |
| ¾ cup | Gruyere cheese, shredded |

Mix the truffle oil, vinegar, salt, pepper, garlic, and shallot in a bowl to make a marinade. Slice the tenderloin at a brief angle into medallions. Coat the medallions with the marinade, lay in a shallow pan, and let them rest while covered in a refrigerator for an hour. Melt butter in a skillet (med. heat) and sauté the pork (conserve the marinade) until it is lightly browned (2-3 min. per side). Transfer to a plate and keep warm. Heat add oil to the skillet and sauté the mushrooms 4-5 min. until they are browned and limp. Add the marinade (including shallot and garlic) and continue to cook until the shallots are soft and translucent and garlic aromatic (2-3 min.). Deglaze with the wine, scraping any fond from the bottom of the skillet. Stir in the stock, any juices on the pork plate, oyster or stir-fry sauce, soy sauce, and Worcestershire sauce, Dijon mustard, and parsley. Cook until the parsley wilts (ca. 2-3 min.). Adjust seasonings (if you find you've added too much truffle oil, try adding a bit of Worcestershire sauce to balance it). Stir in cream and simmer 2-3 min. Stir in flour and simmer until the sauce reaches the desired thickness. Remove from heat and stir in the cheese until it melts. Plate the pork medallions and pour the sauce on top. Serve with mashed potatoes and a salad. Serves 3-4.

---

1    E.g. Laka. 2024.
2    The meat you purchase may come in much larger amounts. This being the case, you might cut everything into medallions, use the appropriate amount for this recipe, and freeze the rest for later use.

# Truffled Salisbury bison steak & mashed potatoes

In 7th grade history class, I was often daydreaming and looking out the window for something more inspiring to think about. The teacher became annoyed one day when I was staring out the window again. In fact, everyone but the teacher was staring and murmuring. He was trying to teach us about how wild buffalo once roamed the prairies not far from our school, and none of us were paying attention. That was because there were actual buffalo grazing on the school's front lawn! Farmer Forbes' domestic herd had broken through a fence, choosing the perfect moment for an historical reenactment. Here in Canada, bison are large-headed, hump-backed relatives of buffalo. They are farmed for meat that you can buy (usually ground) in many B.C. stores. They also still roam free in parts of northern Canada. Once I was driving a truck towards Yellowknife on a very foggy McKenzie Highway, and an entire herd of wood bison suddenly loomed out of the mist. I braked hard. The lead bull took my sudden appearance as a challenge, pawed at the road, snorted, and came right up to the driver's-side window to give me a giant bloodshot hairy eyeball. Talk about anxious! Hopefully, you won't need to be anxious making this hearty bison dish. It makes the perfect meal for a cold winter day.

| Potatoes[1] | | Steaks[2] | | Gravy | |
|---|---|---|---|---|---|
| 5 cups | Water | 320 g | Ground bison | 2 tbsp | Butter |
| 6 | Potatoes, cubed | 1 | Shallot, minced | 2 | Shallots, diced |
| ¼ cup | Milk | 2 cloves | Garlic, minced | ½ lb | Creminis (*A. bisporus*), sliced |
| 1 tbsp | Butter | 1 | Egg | To taste | Salt & pepper |
| ½ tsp | Garlic powder | ¼ cup | Panko breadcrumbs | 1½ cups | Beef stock |
| ½ tsp | Onion powder | ¾ tsp | Italian seasoning | 1 tbsp | Tomato paste |
| 1 tbsp | Parsley flakes | 1 tsp | *Herbs de Provence* | 1 tbsp | Dijon mustard |
| 1 cup | Cheddar cheese, white, shredded | To taste | Salt & pepper | 1 tsp | Worcestershire sauce |
| | | 2 tbsp | Olive oil | 1 tsp | *Herbs de Provence* |
| 1 tsp | Truffle oil | 1 tbsp | Worcestershire sauce | 1 tsp | Seasoned mushroom powder[3] |
| | (*T. melanosporum*) | 1 tsp | Truffle-infused olive oil | 1½ tbsp | Flour |

Bring water to boil in a large pot, add potatoes, and reduce heat to a simmer. Boil potatoes until soft (ca. 20 min.). Meanwhile, mix ground bison, a chopped shallot, garlic, egg, Panko breadcrumbs, Italian seasoning, and *Herbs de Provence* in a bowl. Form into patties. Heat oils in a skillet (med. heat) and cook patties 5-6 min./side, until light brown and cooked entirely through. Remove and keep warm. Melt butter in skillet and sauté remaining shallots until soft and translucent (2-3 min.). Add mushrooms & cook until slices are limp and browned (4-5 min.). Season w/ salt & pepper. Stir in stock, tomato paste, mushroom powder, Worcestershire sauce, and *Herbs de Provence*. Simmer 2-3 min. Stir in flour, and cook until thickened. Drain potatoes, mash with cheese and butter. When those melt, mash in milk, butter, garlic/onion powders, parsley flakes, & truffle oil. Plate patties with mashed potatoes to the side, pour gravy over top. Serve while warm.

---

1   Grace, L. 2022.
2   Dash, T. 2020; Force of Nature. 2020
3   A recipe for this is found in Part 1 of the series (*Beginnings & Springtime*).

Forbes' buffaloes standing in the snow (inside their fence). Painting by Richard Winder

# A Modest Beef Wellington

The origins of Beef Wellington are uncertain. Clearly, it is named for Arthur Wellesley, the first Duke of Wellington and the man commanding the armies that defeated Napoleon at the Battle of Waterloo in 1815. It might be that Wellesley's chef named the dish after him, or that it wrapped things snugly like a Wellington boot, or that the dish was simply renamed from a French dish (the similar *filet de boeuf en croute*). Just to confuse matters, some of the earliest references to the dish appear to be American, for example 'Fillet of beef, *a la* Wellington' appearing in the 1903 *Los Angeles Times*. Julia Child popularized the dish on her TV program *The French Chef*, and here we are![1] The 1965 *New York Times* defined the gold standard for Beef Wellington as a beef tenderloin surrounded by *pâté de foie gras*, truffles, and cognac, all in a pastry crust. Recipes often use duxelles in place of goose liver.

This recipe features duxelles made from *Agaricus bisporus* mushrooms. So, one could group this dish with other recipes involving *Agaricus*. But it also includes the aroma and flavour of truffles (as many Beef Wellington recipes do). Beef Wellington is something you could serve proudly for a winter holiday feast. That makes it a great winter recipe to culminate this series. Given the expense and luxury of the ingredients and involved procedure, one might consider:

> "If it's worth doing, it's worth doing right."
> -Elizabeth Bright (Daman) Braine (1881-1968)

Great Grandma Braine's maxim was famous in the Winder family, and it certainly applies to this famous dish. You'll have to splurge on ingredients, no doubt about it. But 'doing things right' doesn't mean needless spending, working at breakneck speed, or running pell-mell in frantic circles around the kitchen, anxiously seeking perfection in record time. That can be a recipe for frustration, maybe even disaster. In short:

> "This isn't *Iron Chef*. Don't be that person."
> -Chris Young, 2023[2]

Making this dish is relatively straightforward if one plans and patiently follows directions. In practical terms, one will want to pace carefully through the steps and do the preparatory work ahead of time. Take time to plan ahead. Protect the pastry from becoming soggy by making sure that excess juices are cooked or drained and wiped from the meat and squeezed, cooked, and drained from the duxelles. Don't worry about decorating the pastry unless you have time (if you do, be sure to only score it lightly with blunt edge of a rounded knife so that you don't puncture the surface). Respect standing and cooking times. Use a meat thermometer to ensure that the beef cooks to the proper stage. In other words:

> "Don't Panic!!"
> -Douglas Adams, 1978[3]

What makes this a recipe for a 'modest' Wellington? Well, an average entire beef tenderloin cut will likely weigh between 4 to 6 pounds (1.8 to 2.7 kg). Most Wellington recipes that you will find online settle on an amount that's about half of that (2 to 3 pounds or 0.9 to 1.3 kg). That's still quite a costly quantity, especially when you trim the tapered cut to be more evenly cylindrical. Most local grocers here in B.C. tend to sell beef tenderloin in cuts of ca. 0.25 kg (although you can order larger cuts from the butcher). Tenderloin is often sold in the form of fillets or medallions that may or may not be labelled as fillet mignon. I decided to combine two fillets for my Wellington, amounting to about 0.5 kg in total. The result is therefore relatively small, and cooking temperatures and times have to be adjusted accordingly. But doing things this way keeps things relatively affordable. Perhaps you can afford a larger portion. If so, you may need to increase the amount of duxelles that you prepare. Also, be sure to scope out information on cooking times and temperatures you will need for that amount and your desired level of doneness. If you're still unsure, and you have some porcini on hand, consider trying the Petite Beef Wellington recipe found in Part 3 as a test run. It uses ground beef and may be a bit more forgiving (and less expensive). It also uses phyllo pastry sheets. Phyllo pastry seals filling liquids without the pastry becoming too soggy. But if you want to plunge right in to this one, don't worry! It will turn out just fine and your dining partner(s) will truly enjoy it and be amazed. *Bonne chance et bonne soirée*!

---

1    Gordon Ramsay Restaurants. 2022.
2    Young, C. 2023.
3    Adams, D. 1978.

## Duxelles

| | |
|---|---|
| 1 tbsp | Olive oil |
| 1 | Shallot, chopped fine |
| 1 lb | Cremini mushrooms (*Agaricus bisporus*) |
| 5 sprigs | Thyme leaves, minced |
| 1 tsp | Truffle oil (*T. melanosporum*) |

## Mushroom sauce

| | |
|---|---|
| 2 cups | Mushroom broth |
| 2 tsp | Ketchup |
| 1 tsp | Dijon mustard |
| 3 tbsp | Red wine |
| ½ tbsp | Worcestershire sauce |
| 1 tsp | Summer savory |
| ½ tsp | Soy sauce |
| ½ tsp | Onion powder |
| ½ tsp | Garlic powder |
| 1 tsp | Balsamic vinegar |
| 1 tbsp | Miso paste |
| 1 tbsp | Oyster sauce |
| 1 tbsp | Black bean sauce |
| 1½ tbsp | Blackberry-flavoured vinegar |

## Beef & pastry

| | |
|---|---|
| 0.5 kg | Beef tenderloin |
| 1 tsp | Thyme leaves, minced |
| ½ cup | Mustard, Dijon[1] |
| 1 tbsp | Mustard, yellow |
| To taste | Salt & pepper |
| 400 g | Puff pastry |
| 3 layers | Phyllo pastry |
| 3 tbsp | Beef fat[2] or butter |
| 1 | Egg, beaten |

---

1 Typically there is mustard in Wellington recipes, but the reccomended type varies. Since the origins of this dish are uncertain (maybe English or possibly French), there isn't a mandatory 'traditional' mustard choice in this author's humble opinion. An English format would use yellow mustard, a French format would use Dijon mustard (better tasting according to some). A North American format might use either or both (as in this recipe). Use whatever makes your heart content and your taste buds happy.

2 Fat drained from cooking beef hamburgers and stored in congealed form in a small jar.

**Preparing the duxelles:** Prepare duxelles a day ahead. Finely chop mushrooms and shallots. Heat oil in skillet (medium heat) and sauté shallots 1-2 min. until soft and translucent. Add the mushrooms and continue to sauté until the mushrooms are browned and well caramelized (perhaps 5-6 min.) Stir in thyme, cook 1-2 min., and remove from heat. Cool and use a colander/strainer stand and pestle to force all excess liquids from the duxelles. Freeze duxelles, thaw and add truffle oil when ready to use.

**Preparing a mushroom sauce:** If you are serving the Wellington with gravy or sauce, it would be a good idea to also prepare this ahead of time (e.g. when you prepare the duxelles). You could use regular beef gravy, or consult part 1 of this series (which has a number of sauces that you could consider). Alternatively, add mushroom broth to any liquid you extract from the duxelles, bringing the total volume up to 2 cups. Add ketchup, Dijon mustard, red wine, Worcestershire sauce, summer savory, soy sauce, onion powder, garlic powder, balsamic vinegar, tsp miso paste, oyster sauce, black bean sauce, and blackberry-flavoured vinegar. Bring to a simmer and melt 2 tbsp butter in the sauce. Gradually stir in ¼ cup flour so that it doesn't clump, and simmer until thickened. Let it cool and store in a jar. Refrigerate or freeze until baking day (if freezing, don't fill the jar up all the way, and try to use one with straight sides rather than shoulders.

**Preparing the tenderloin:** On the day before baking the Wellington, pat the beef dry with a paper towel and use a sharp knife to make a cross-hatched diamond pattern of surface scratches. Sprinkle the thyme, salt and pepper onto the meat. Heat the beef fat or butter in a skillet (medium heat) and lightly sear each side of the beef as well as the ends. Remove the skillet and wipe away liquids with a paper towel when safe to handle). Mix the mustards and spread a very thin coating on the surface of the meat, wrap it tightly in plastic wrap, and place it in a refrigerator (5°C) overnight. If package directions indicate it, also move the puff and phyllo pastries from the freezer to the refrigerator so that it will thaw while remaining chilled.

**Making & baking the Wellington:** On baking day, preheat oven to 400°F and prepare a clean working surface. Use a rolling pin to roll out the chilled puff pastry as indicated in package directions, spreading out enough to wrap around the meat. Lay a 3-ply layer of phyllo sheets on top of the puff pastry, trimmed to fit the puff pastry. Spread a layer of duxelles onto the phyllo pastry (not making it too thick, and leaving 1" margins around the sides bare). Wipe excess mustard from the meat and lay it onto the duxelles layer (in the centre of the pastry). This dish was tested with two 0.25 kg medallions that were stacked.[1] Season with salt and pepper, and fold the pastry around the beef (moistening the edges with a bit of the beaten egg if necessary and pressing to seal the edges). Optional: use the blunt edge of a rounded knife to very lightly score the surface of the pastry with a decorative pattern (without puncturing- not shown). Alternatively, you might also cut strips from the some of the remaining pastry and use the egg wash to stick them to the main pastry in a pleasing decorative pattern of your choice (also not shown). Use the egg wash to brush the outer surface of the pastry. Use a small knife to make several discrete punctures that will allow for some venting. Place on a tray or sheet and bake in the oven to desired level of doneness—the pastry should be crisp and the surface should turn a golden colour. If using 0.5 kg (1 pound) of tenderloin cook for 30 to 40 minutes to reach medium rare, or 45 to 50 minutes for medium. The Wellington tested here was baked for 45 min. and reached a minimum of 160°F when removed from the oven.

**Serving the Wellington:** After removing the Wellington from the oven, the internal temperature (measured with a meat thermometer) should reach about 135t o 145°F (57°C to 63°C) for medium rare, 150-160°F (65.5°C to 71°C) for medium, and 170°F (77°C) for well done. Allow the pastry to rest a further 40 min. (during which time it will continue to bake internally). Slice and plate with your favourite mushroom sauce while still warm. Serve with your best mashed potato recipe. Serves 4-5.

---

1    This approach doesn't result in a single perfectly round section of meat in the centre of each slice of the Wellington, but it is a convenient alternative.

# useful adjuncts

# Olive oil

Olive oil mainly consists of mono-unsaturated fatty acids that are better for your health. Unlike seed and nut oils, olive oil is actually a fruit juice (yes, olives are a botanical fruit—a very oily one). That means that its flavour will degrade if stored too long or improperly. Keep it away from the heat and direct sunlight of a kitchen counter, and instead store it in a well-sealed (preferably opaque) bottle, in a cool, dark cabinet. The shelf-life of a sealed bottle is usually 2 years after harvest. When opened, a bottle will probably last 2-3 months before gradually degrading and acquiring a waxy, eventually rancid taste. Purchasing smaller amounts and using it regularly avoids spoilage and lets you try out all of the different types.[1]

Olive oils from different regions and cultivars have distinctly different flavours (Table 8), although they are often blended. Sometimes you see bottles of olive oil emblazoned with the term 'cold-pressed' or 'cold-extracted.' It is a marketing term. In ancient times olive oil was made by crushing the olives with millstones and pressing the paste between burlap mats. Virtually all commercial oils use more modern cold-pressing methods. The olives are crushed and churned by machines at a mill to release the oil and let it coalesce. In order to prevent degradation by heat, the temperature of the churning paste is not allowed to exceed 80.6°F. The paste is then centrifuged to separate the oil from the pumice. A more important distinction is quality and the level of refinement. Refined oil is treated to remove impurities and raise the smoking point, but has less flavour and lacks the antioxidants, nutrients, etc. of unrefined oils. Extra-virgin and virgin olive oils are unrefined. They have lower smoking points, but full flavor and nutrient content, and no more than 1% oleic acid. The difference between the virgin and extra-virgin amounts to the level of standards in processing, and slightly higher levels of oleic acid in the virgin type. 'Pure' or regular olive oil is normally a blend of virgin and refined oil.[2] Olives used for extra-virgin oils are harvested during a narrow window of time when the fruit is just ripe enough to have the best, intense flavor and optimal oil content, before the taste mellows and oil increases with further maturation. Given that the olives are more difficult to shake loose and have less oil at this stage, and must move from field to mill in 3 hours or less to avoid degradation in the heat of the day, it is a frenzied harvest.[3]

The cost of extra-virgin olive oil clearly reflects the higher expense of production, but can you be sure you're getting your money's worth? Fraud has been a feature of olive oil markets since the days of the Roman Empire. In the contemporary market, you might find olive oil blended with holdover oil from previous years, or low-grade olive oil diluted with soy oil or seed oils. You will see all sorts of tests recommended online, ranging from heating it or seeing if it will burn, to complicated chemical analysis. At the end of the day, the best way is simply to taste it. You might have paid a lot for it, but if your extra-virgin oil lacks flavour, falls flat, or tastes off or rancid, then raise an eyebrow. Of course, you can't just sip olive oil out of all of the different bottles at the store and taste them all before you choose. Look for virgin or extra-virgin oils that have a harvest date and provenance (best-before dates are not the same thing). Also look for certifications like Europe's PDO, Italy's DOP, or California's COOC, or for oils from Australia and Chile where standards and testing are stringent.[4]

1    Jolly, N. 2020b.
2    Foster, K. 2015.
3    Jolly, N. 2020a,b.
4    Fuller, J. 2017.

Table 8. Source and flavours of some notable varieties of olive oil.[1]

| Cultivar | Region | Flavour |
|---|---|---|
| | Italy | |
| | (North) | |
| Lavagnina | Liguria | Very light and delicate, with smooth buttery finish |
| Pignola | Liguria | Smooth, buttery, fruity, stronger flavor if grown at colder temperatures |
| Taggiasca | Liguria | Delicate, buttery, fruity, with small notes of bitterness |
| Razzola | Liguria | Delicate, herbs, buttery |
| | (Central) | |
| Frantoio | Tuscany, etc.[2] | Intense, especially artichoke leaf, fresh cut grass, green almonds[3] |
| Gentile di Chietti | Abruzzo | Delicate, balanced, elegant, herbaceous (cut grass, green almonds), spicy finish |
| Leccino | Toscana[4] | Delicate, captivating, cut grass, almond, lightly spicy ending |
| Moraiolo | Umbria, Toscana | Bitter, spicy, herbaceous, green almond, artichoke, tomato, aroma fruity, floral |
| Olivastra Seggianese | Toscana | Delicate, olive fruit aroma; creamy texture, buttery, nutty taste, peppery hint |
| Pendolino | Lazio, Toscana | Herbal aroma, notes of almonds, spicy ending |
| | (South) | |
| Biancolilla | Sicilia | Light olive, slightly bitter, hints of artichoke, light spicy finish |
| Bosana | Sardegna | Satisfying, fruity, white fruit, bananas, apples, delicate olive aroma |
| Cerasuola | Sicilia | Medium fruitiness, very intense, bitter & spicy, notes of olive, hay |
| Coratina | Puglia | Elegant, spicy, a little bitterness, adds character to blends |
| Giraffa | Sicilia | Lightly fruity, herbaceous, green without spice |
| Minucciola | Campania | Bright aromatic, notes of apple, wildflowers; well-balanced bitter, spicy notes |
| Moresca | Sicilia | Well-rounded, elegant notes of artichoke, aromatic herbs, greens; spicy finish |
| Nocellara della V-B.[4] | Sicilia | Vibrant, robust, grassy, herbaceous, very spicy |
| Nocellata dell'Etna | Sicilia | Fruity, green tomato, tomato leaf, artichoke, almond, spicy & light bitter finish. |
| Ogliarola | Puglia, etc.[5] | Deeply fruity, (apple, white peach), grassy, light almond aftertaste, spicy finish. |
| Ortolana | Campania | Elegant, soft, fruity sensations, spicy |
| Ortice | Campania | Spicy, slightly bitter, well-balanced tones of fresh tomatoes and vegetables |
| Ottobratico | Calabria | Balanced herbaceous, artichoke, thistle, sour apple, sweet almond, spicy finish |
| Peranzana | Puglia | Freshly cut grass, sweet, fruity like artichoke, lingering spiciness |
| Recioppella | Campania | Very aromatic, spicy, bitter with notes of fresh herbs, basil, mint, sage |
| Tonda Iblia | Sicilia | Fruity, full-bodied, spicy, notes of green tomato, fresh herbs |

(Continued on next page.)

---

1   Eataly. 2015.; Paulina. 2012; Olive oils from Spain. 2022.
2   Also Umbria, Lazio
3   Lingering bitter harmony and peppery aftertaste
4   Nocellara della Valle de Belice
5   Also Basilicata, Campania

Table 8 (continued). Source and flavours of some notable varieties of olive oil.[1]

| Cultivar | Region | Flavour |
|---|---|---|
| | Greece | |
| Kalamon (Kalamata) | Messinia, Lakonia, Lamia | Rich and fruity |
| Konservolia | Amfissa, etc.[2] | Mellow, pleasurably sweet |
| Koroneiki | Crete | Robust but mild, fresh and fruity, pungent, pleasant, peppery finish |
| Ladolia (Kutsurelia) | Pelopennese | Grassy, herbal |
| Thasitiki (Throumba) | Thassos | Elegant, complex, tomato leaf, dried herbs, spices |
| | Spain | |
| Alfafara | Alfafara (Alicante), etc.[3] | Balanced bitterness and spiciness with banana and ripe-tomato notes |
| Aloreña | Málaga | Green fruit, grass |
| Arbequina | Catalonia, etc.[4] | Fluid & sweet, fruity (olive, apple, banana, almond) |
| Blanqueta | Alicante, etc.[5] | Intense fruitiness, strong peppery and bitter taste |
| Castellana | Guadalajara, etc.[6] | Fruity, spicy, very aromatic, notes of green grass, leaves, banana |
| Changlot Real | Valencia | Green notes with hints of apple, wood, and almond |
| Cornicabra | Toledo, etc.[7] | Aromatic, fruity, olive, herbal olive leaf, apple & other fruits, slightly bitter |
| Empeltre | Aragón, etc.[8] | Sweet, intensely fruity (green apple), almond, slightly peppery and bitter |
| Farga | Castellón, etc.[9] | Intense flavor, recalls flavours of almond and walnut |
| Hojiblanca | Málaga, etc.[10] | Sweet foretaste, cut grass, artichoke, aromatic, slight bitterness, peppery finish |
| Lechín de Sevilla | Seville, etc.[11] | Very fruity, balanced green, peppery and bitter notes |
| Manzanilla Cacereña | Badajoz, etc.[12] | Peppery, bitter, sweet, intensely fruity, grass, green apple, fruit salad, tomato |
| Morisca | Badajoz, northern Seville | Very fruity, sweet |
| Morrut | Baix Ebre-Montsía | Spicy, very fruity, slightly bitter, almond and green apple flavours. |
| Picual | Jaén, etc.[13] | Full bodied, fruity, green olive, light peppery and bitter olive leaf[14] |
| Picodo | Córdoba, etc.[15] | Highly aromatic, green fruity flavor with green apple, olive leaf notes |
| Royal | Northern Jaén | Sweet, very fruity with dominant fig notes |
| Sevillenca | Tarragona, Castellón | Very fruity, sweet, peppery with slight bitterness, green apple and banana notes |
| Verdial from Badajoz | Badajoz, Cáceres | Slightly sweet and bitter, tones of almond and ripe fruit. |
| Verdial from Huévar | Huelva, Seville, Cádiz | Fruity, intensely spicy and bitter, green tones, green almond and figs. |
| Verdial de Vélez-Málaga | Malaga | A bit peppery, intense taste of ripe fruit. |
| Villalonga | Valencia, Alicante | Intensely fruity, slightly spicy & bitter, notes of cut grass, almond, apple |

---

1   Eataly. 2015; Paulina. 2012; Olive oils from Spain. 2022.
2   Also Volos, Agrinio, Atalanti, and elsewhere
3   Also Valencia, Albacete
4   Also Alto Aragon, and elsewhere
5   Also Valencia, Murcia, Albacete
6   Also Cuenca and Madrid
7   Also Cuidad Real, Madrid, Cáceres, Badajoz
8   Also Balearic Islands Castellón, Tarragona, Navarr
9   Also Tarragona, Lérida
10  Also Córdoba, Granada, Seville
11  Also Córdoba, Cádiz, Málaga
12  Also Cáceres, Salamanca, Ávila, Madrid
13  Also Córdoba, Granada, Castilla-La Mancha, and elsewhere
14  Prized for stability (resistance to oxidation, high temperatures).
15  Also Granada, Málaga, Jaén

# Chilis

Chilis (*chili* = American/Canadian, *chile* = Spanish, *chilli* = Oxford English, Náhuatl/Aztec) are important ingredients in many recipes throughout this series. Too much spice can overwhelm a savory mushroom dish, but a tiny bit can perk up otherwise pedestrian fare. Chilis are one of the oldest cultivated crops in the Americas starting in east-central Mexico six millennia ago. Italian explorer/exploiter Christopher Columbus called them 'peppers' because their spiciness reminded him of black pepper (*Piper* spp.). Following their discovery by Europeans, chilis were carried all around the world and developed into many regional varieties (Table 9). All chilis are *Capsicum* spp., with domesticated species spreading from the Americas to global fame and adoration: *C. annuum*, *C. frutescens*, *C. chinense*, *C. pubescens*, and *C. baccatum*. Capsaicin (8-methyl-*N*-vanillyl-6-nonenamide) is the compound providing their spicy heat, gauged on a relative scale of 'Scoville units.' There is a psychological benefit to that heat (enjoyment of constrained risk), as well as a complimentary taste benefit that tends to bring out more from some flavours (or at least wake them up).

Table 9. Some chili/pepper varieties and their characteristics

| Variety | Species | Origin | Heat(1,000 SU)[1] |
|---|---|---|---|
| Bell (sweet) | *C. annuum* | N. & S. America | 0 |
| Banana pepper | *C. annuum* | South America | 0-0.5 |
| Friggitello (Sweet Italian pepper) | *C. annuum* | Americas to Italy | 0.1-0.5 |
| Poblano (dried = ancho) | *C. annuum* | Mexico (Puebla) | 1-2 |
| Kashmiri red (Kashmiri lal mirch) | *C. annuum* | Americas to India | 1-2 |
| Korean | *C. annuum* | Americas to Japan to Korea | 1.5 |
| Padrón | *C. annuum* | Americas to Spain (Galicia) | 0.5-2.5 |
| Anaheim | *C. annum* | New Mexico to California | 0.5-2.5 |
| Hungarian wax pepper/paprika | *C. annuum* | Americas to Hungary | 1-15 |
| Fresno | *C. annuum* | California | 3.5-10 |
| Jalapeño (ripe, smoke-dried = chipotle) | *C. annuum* | Mexico | 3.5-10 |
| Aleppo (Halep biberi, Halaby) | *C. annuum* | Americas to Syria, Turkey | 10 |
| Cheonyang (Korean pepper) | *C. annuum*[2] | Americas to Asia to Korea | 10 |
| Serrano | *C. annuum* | Mexico (Puebla, Hildalgo) | 10-23 |
| Hot Portugal | *C. annuum* | Americas to Portugal (U.S.?) | 5-30 |
| Peperoncino | *C. annuum*[3], *C. frutescens*[4] | Americas, then Italy | 15-30 |
| New Mexico (Green) | *C. annuum* New Mexico group | New Mexico | 0-70 |
| Japones (hotanka, santanka) | *C. annuum* | Mexico (Jalisco) to Asia, etc. | 15-30 |
| Aji colorado | *C. baccatum* | Peru | 20-30 |
| Aji amarillo | *C. baccatum* | Peru | 30-50 |
| Cayenne | *C. annuum* | Brazil | 30-50 |
| Tobasco | *C. frutescens* | Mexico | 30-50 |
| Chiltepin - pequin | *C. annuum* var. *glabriusculum* | N. & S. America | 30-50 |
| Yatsufusa | *C. annuum* | Americas to Japan | 40-75 |
| Tien-tsin (Chinese red pepper) | *C. anuum* | Americas to China | 50-75 |
| Chiltepin - tepin | *C. annuum* var. *glabriusculum* | N. & S. America | 50-163 |
| Bird's eye (de arbol, Thai, etc.) | *C. annuum* | Mexico to various | 50-100 |
| Peri-peri | *C. frutescens* | S. America to S. Africa | 50-175 |
| Rocoto (locoto, manzano) | *C. pubescens* | N. & S. America | 2.4-250 |
| Guntur | *C. chinense* | N. & S. America to India | 30-350 |
| Habanero | *C. chinense* | Amazon to Mexico | 100-350 |
| Scotch Bonnet | *C. chinense* | Caribbean | 100-350 |
| Bhut jolokia (Bhutan/ghost[5] pepper) | *C. chinense* X *C. frutescens* | Americas to Bhutan & India | 855-1,041 |
| Carolina reaper[6] | *C. chinense* X *C. frutescens*[7] | United States | 1,641 |

---

1　1,000s of Scoville Units
2　Korean pepper X Thai pepper
3　Cayenne
4　Tobasco
5　'Ghost' is a mistranslation.
6　Do not consume this one raw, no matter what challenge your read about on the internet. The capsaicin levels of this variety are actually hazardous.
7　Bhut jolokia X habanero

# Tomatoes

If we are going to elevate our cooking with mushrooms, we should also know something about the things we frequently cook with them (like tomatoes). Tomatoes are a fascinating concept. Originally from the Andes Mountains of Peru and Ecuador, they were domesticated by the plant-breeding geniuses of Pre-Columbian Mexico (where the Aztecs called the plant *tomatl*). The plants produce a berry-type fruit (botanically speaking) that people more conventionally classify as a vegetable. The popularity of tomatoes grew like an indeterminate vine, albeit slowly at first. When Spanish explorers brought them back to their 'Old World', Europeans couldn't make up their mind about them. Plants introduced to Italy from Morocco were called *pomi di mori* (apples of the Moors), and when they made their way to France they became *pommes d'amour* (apples of love). Was that a corruption of the Italian name, or did they have, as some claimed, aphrodisiac properties? Given an overall propensity for frequently claiming that things had aphrodisiac properties, one suspects that people of the day were quite impressionable. They do look jolly and red on the outside, but their slimy, seedy interior is a real turn-off for some. Other Europeans simply didn't trust them, suspecting that they could be a poisonous relative of deadly nightshade (actually, they were correct in that both belong the plant family we call Solanaceae, along with potatoes, ground cherries, tomatillos, tobacco, etc.). Beyond their tart, sweet, umami taste and vitamin C they are also noted for producing lycopene, a very red carotene-like hydrocarbon. But the story doesn't end there. Consumption of tomatoes continued to grow throughout Central and South America (the land of a million salsa recipes), Europe and particularly Italy (the land of a million pasta sauce recipes), and eventually North America (the land of a million ways to pour ketchup on things). They eventually became so popular that they elbowed mushrooms out of the ketchup spotlight,[1] and created further controversy when ketchup made with tomatoes was tentatively discussed as a potential 'vegetable' category for school lunches in the 1980s in the United States. This weird and wonderful vegetable concept finds new heights of weirdness and wonder in a certain country's fondness for tomato ketchup-flavoured potato chips (we're looking at *you*, Canada).

It's well worth matching the variety of tomato or tomato sauce to the type of recipe. For example, pasta sauces often benefit from the use of passata versus other types of commercial tomato sauce. Passata is a tomato purée that is more processed than a whole, peeled tomato, but doesn't have a lot of the cooking with spices, oil and salt of commercial ready-to-heat sauces. As such then, passata has a brighter, bolder, fresher flavour and gives you, the cook, more control over the final flavour in a dish.

When it comes to types of tomatoes, many strange and fantastic varieties have been developed after their domestication. Some of the general types include cherry and plum tomatoes (among of the best to use for flavour in all kinds of dishes), pretty pear-shaped varieties, varieties with different colours (red, scarlet, orange, yellow, and even purple), big beefsteak tomatoes, determinate (bushy) and indeterminate (staking, vine-like) varieties, disease-resistant varieties, cold climate and early-bearing varieties, and oblong varieties used for making sauce. You often see Roma tomatoes recommended in recipes, which is not really surprising since it is a plum tomato often sold in supermarkets. Some very old heirloom varieties like deeply ribbed 'Purple Calabash' haven't changed much since domestication and have fantastic flavour, with an aroma that permeates the entire kitchen when you slice one open. Organic farmers often offer heirloom varieties that have tons of flavour compared to the run-of-the-mill types grown in larger commercial operations. It's well worth supporting their business if you are taking the time to cook mushroom dishes that emphasize umami richness. Plant breeders in Canada have produced many Canadian varieties over the years, and it's also worth exploring the diversity that they offer for the table (especially if you're a gardener).[2]

This discussion couldn't be closed without mentioning a Canadian tomato variety that greatly appeals to my inner botanist: 'Quebec #13.' The antithesis of Purple Calabash, Quebec #13 is a variety that can inspire plant breeders everywhere with the sheer audacity of its intentional design. True, it tastes like cardboard—but that means it addresses the widest possible market. The watery, bland, nondescript flavour plays to the lowest common denominator, avoiding the notice of picky eaters and milquetoast consumers alike. But you have to look beyond that, because it also has other outstanding features. Quebec #13 produces an excessively sturdy and rugged berry (tougher than the cardboard taste would suggest). It positively shines when it comes to its very firm, robust texture. That kind of durability comes in handy when you've been dumped into large shipping containers and then jostled over long distances in speeding semi-

---

1    See discussion of mushroom ketchup in Part 1.
2    Rafuse, C. 1990.

trailers. But wait! Let's not forget the *pièce de résistance*: its fabulous, approximately cubical shape (with slightly rounded corners), providing optimal packing density for efficient, cost-effective transport and distribution. You can hardly find another variety like it![1] All kidding aside, if you don't have a garden look for vine-ripened tomatoes when you visit the store. Roma types are usually a good bet. You might pay more, but this is one case where quality makes a difference.

Table 10. Some tomato varieties offered and/or grown or reccomended in B.C., Canada and their characteristics.

| Variety | Fruit | Plant type[2] | Resistance[3] | Description |
| --- | --- | --- | --- | --- |
| Acclaim | Round | Det., Hybrid | ? | Commercial tomato, like 'Jackpot' |
| Amish Paste | Large plum | Indet., Open | (Good) | Long season, meaty, heirloom for canning |
| Better Boy | Large red, 1lb | Indet., Hybrid | V,F,S,N,A | Delicious, crisp, juicy, deep red, slicing |
| Black Krim | Large purplish | Indet., Open | (General) | Ukraine heirloom, beefsteak flavour |
| Bonny Best | Med. red round | Indet., Open | (Good) | Super taste, uniform, old-fashioned heirloom |
| Bush Beefsteak | Large flat globe | Det., Open | ? | Early, juicy, meaty heirloom |
| Celebrity | Med. deep oblate | ~Det., Hybrid | V,F,N,T,A | Consistent, flavourful |
| Chocolate cherry | Brick red cherry | Indet., Open | ? | Heirloom, delicious complex flavour |
| Cupid | Scarlet Grape | Indet.Hybrid | F,S,A | Bright colour, sweet, flavourful |
| Early Girl | Globe | Indet., Hybrid | F | Early bearing, prolific |
| Fantastic | Globe | Indet., Hybrid | V,F,N | Juicy, meaty, beefsteak flavour |
| Florida 47 | Large oblate | ~Det., Hybrid | V,F | Firm, vine ripe |
| Gardner's delight | Cherry | Indet., Open | (General) | Bright red, heirloom flavour, needs heat |
| Honey Bunch Red | Red grape | Indet., Hybrid | A | Delicious, sweet, crack resist., heavy bearing |
| Jubilee | O.-yellow round | Indet., Open | A | Heirloom, delicate, sweet, mild flavour |
| Manitoba | Red globe | Det., Open | V,F | Heirloom, early, fresh, tangy, like Scotia |
| Mortgage Lifter | Pink ribbed | Indet., Open | V,F,N | Beefsteak flavour, heirloom, crack resistant |
| Mountain Crest | Large fruit | Det. Hybrid | V,F | Extended shelf-life |
| Mountain Fresh | Deep oblate | Det., Hybrid | V,F,N | Cool/wet climate, sweet, firm, flavourful |
| New Yorker | Orng.-red globe | Det., Open | V,A,P | Early, cool conditions, heirloom beefsteak |
| Old German | Large yellow/red | Indet., Open | (Good) | Mennonite heirlm., 2-coloured, sweet slicing |
| Patio | Round red | Det., Hybrid | F,S,A | Dwarf, containers & baskets, juicy, delicious |
| Pink Brandywine | Pink ribbed | Indet., Open | ~ | Fruity, aromatic heirlm., full sun/greenhouse |
| Purple Calabash | Purple ruffled | Indet., Open | (V. Good) | Intense, tangy heirlm. flavour, drought resist. |
| Pruden's Purple | Large, dark pink | Indet., Open | (Good) | Delicious heirloom, beefsteak, full sun |
| Quali T 23 | Flat globe | Det., Hybrid | VFT | Late season market tomato |
| Roma | Plum | Det., Open | V,F | Firm, slender, good for sauces, canning |
| San Marzano | Elongate plum | Indet., Open | (Good) | Ital. heirloom, Incredible sauce, flavour |
| Scotia | Globe | Det., Open | (Good) | Cold tolerant, general use & greenhouse |
| Smarty | Grape | Indet., Hybrid | V,F | Firm, uniform, good taste, high sugars |
| Stupice | Small 2" | Indet., Self | P | Cold tolerant, reliable, sweet and tart |
| Sun Brite | Large oblate | ~Det., Hybrid | V,F,S,A | Adapted for stake or ground culture |
| Sun Chief | Standard | Det., Hybrid | V,F | Early, firm fruit |
| Sun Sugar | Gold cherry | Indet., Hybrid | T | Sweet, golden colour |
| Super Sweet 100 | Cherry | Indet., Hybrid | V,F | Super sweet, like the name says |
| Sungold | Orange cherry | Indet., Hybrid | V,F,T | Orange-coloured, sweet and tart |
| Sweetie | Red cherry | Indet., Hybrid | A | High sugar, juice/preserves |
| Sweet Million | Cherry | Indet., Hybrid | F,T | Vigorous, hardy, delicious |
| Ultra Boy | Large round | Det., Hybrid | V,F,N | Old-fashioned flavour |
| Zapotec | Pinkish pleated | Indet., Open | (General) | Sweet flavour, heirloom, drought resistant |

---

1    Please, just don't use it in any recipes.

2    I = Indeterminate, Det. = Determinate, ~Det. = Semi-determinate, Open= Open-pollinated, Hybrid = Hybrid cross

3    Disease resistance. V = *Verticillium*, F = *Fusarium*, T = Tobacco mosaic virus, P = *Phytophthora infestans* (late blight), A= *Alternaria* (stem canker), S = *Stemphylium* (grey leaf spot), N = Nematodes

# Home-made wines

Yours truly no longer drinks alcohol, but wine-making is still a hobby that can be both fun (you can make your own brand labels!) and useful when it comes to cooking. It's certainly handy to have a selection home-made wines on hand when cooking with mushrooms. That includes wines that purists would generically classify as hedgerow, fruit, or country wine.[1] Even when you cook away the alcohol, the flavour of plum wines nicely complements chanterelles. Blackberry wine makes all the difference in the flavour of mushroom ketchup.[2] As you learn about wine-making, you can also enjoy learning more about the life of the fabulous fungal yeasts that provide the elixir. Alas, there really isn't room here to fully explore their biology. On top of that there are a lot of tricks to the art worthy of a dedicated book all of its own. There *are* lots of books and online articles out there that can help one learn the craft of wine-making. The wine recipes included here are inspired by those found in Acton & Duncan's 1992 guide.[3] The assumption made here is that you are already familiar with basic wine-making, are familiar with the terminology, have the equipment and supplies, and are ready to stock your cellar with some useful wines for cooking mushrooms.

## Blackberry port

| | |
|---|---|
| 27 lb. | Armenian blackberries[4] |
| 4.4 lb. | Bananas |
| 8.8 lb. | Raisins |
| 1 tbsp | Yeast nutrients |
| 1 packet | Sherry yeast |
| 450 g | Sugar |
| <20 litres | Water |

Peel bananas (reject skins), slice and boil in 13.2 L water for ½ hour. This can be done in 3-L (6 pt.) batches. Strain over crushed blackberries and raisins, add yeast nutrients. When cool pitch yeast, ferment 4-6 days. Strain off, press pulp lightly. Add sugar in stages: 600 mL sugar syrup (450g sugar / 1 L water) when the SG drops below 5, until fermentation ceases. Alcohol content exceeds 20%. Rack into secondary (conserve pulp for second wines). Wine should be tawny and drinkable within a few months.

## Sauternes-style[5]

| | |
|---|---|
| 14.9 kg | Yellow plums, pitted,[6] & 5 kg bananas |
| 5 kg | Bananas |
| 3.1 L | White grape concentrate[7] |
| 1.5 L | Rose petals[8] |
| 82.5 g ea. | Tartaric and malic acids |
| 247.5 ml | Glycerol (Glycerin) |
| 3.7 kg | Sugar |
| 1 pack | Sauternes yeast[9] |
| Some | Yeast nutrients (per directions) |
| Some | Pectic enzyme (per directions) |
| ≤ 21 L | Water |

Peel bananas (reject skins), slice and boil in 19.25 litres water. Strain carefully over the pitted plums. Add acids and nutrients while hot & dissolve the sugar in the hot liquor. When cool, add the rose petals (or oil) & yeast. Ferment on the pulp 3-4 days, strain off pulp and press lightly (or siphon off if making a second wine). Add grape concentrate and glycerol, dilute to 5.5 gal with water. Sugar additions continue in 1¼ lb increments as a syrup (use ¼ lb sugar in 150 ml water). Wait until fermentation ceases before adding the next portion; stop once fermentation stops. Rack wine off the lees in 4 months, earlier if there is a deposit. Rack at 4 month intervals, topping up with water each time. May be drinkable at 6 months, peak at 2 years.

---

1   As opposed to wine made from grapes. Purists are always raining on everyone's parade, aren't they?

2   Keep in mind that wine can be overdone. The best use of wine in cooking enhances flavours without overwhelming them. Three tablespoons may provide enough subtle influence for many sauces. Recipes that want you to use it more generously may be emphasizing a liking for the flavour of wine more than a liking for the overall dish.

3   Acton, B. and Duncan, P. 1992.

4   The invasive species found on Vancouver Island is *Rubus armeniacus*, sometimes called Himalyan blackberry despite its original Armenian distribution. Fresh berries make wine with a fresh, fruity taste. Frozen berries provide a wine that a bit more mellow, rich, and complex in flavour, but also watch out for foaming during secondary fermentation.

5   This will produce a relatively sweet wine with over 20% alcohol content.

6   Ensure that they are European type plums. You can also use yellow cherry-plums (mirabelle plums). Japanese yellow plums are very watery and make a wine that is insipid with a strange, bitter aftertaste.

7   You can use Welch's frozen white grape juice concentrate in place of more formal brewing products.

8   In place of yellow rose petals you can use essential rose oil. In concentrated form, no more than 2-3 drops.

9   You can substitute a champagne yeast (*Saccharomyces cerevisiae* var. *bayanus*) for the Sauternes yeast. If you do, beware that the wine will be quite carbonated and should probably stored in 1-gallon jugs with fermentation locks.

# Latin sausages

Depending on the location outside of Europe, it can sometimes be difficult for some to find a ready supply of some regional sausages (particularly when you need it right away for a mushroom recipe). This section is here to help with two approximate alternatives.

### Italian sausage

There are a wide variety of regional Italian sausages (*salciccia*). This recipe can't stand in for all of them, but it should work very well as a general sausage to be crumbled during cooking, suitable for toppings (hint: pizza!) and sauces (e.g. pork pasta on p. 66 or *Pasta alla Norcina* in Part 2).

| | | |
|---|---|---|
| ½ lb | Ground pork, refrigerated (5°C) or frozen and just thawed | Add all of the ingredients to a bowl and mix them with a clean spoon or spatula (soak your utensils a few moments in clean hot water to be sure). Try to press, smear, and stretch out the meat during mixing, until everything is a bit tacky to the touch. Cover with cellophane, place back in the refrigerator (5°C) and let the sausage rest for 1-2 hours before browning it. |
| 1-2 cloves | Garlic, minced | |
| 1½ tsp | Salt (kosher, sea, or pickling) | |
| 1½ tsp | Black pepper, ground | |
| ½ tsp | Nutmeg | |
| 1½ tsp | Rosemary | |
| 2 tbsp | Wine, white, Italian | |
| 1 tbsp | Olive oil | |
| ¼ tsp | Baking soda | |

### *Chorizo*

In Spain, *chorizo* is made from coarsely-chopped (sometimes ground) pork and pork fat seasoned with garlic, *pimentón* (smoked paprika), salt, wine, and perhaps other herbs and ingredients. There are many regional varieties, but generally the meat is fermented, cured, and smoked before being wrapped in a casing. Mexican chorizo also comes in many regional varieties, and is generally spicier than Spanish chorizo. It may use different types of ground meat rather than chopped pork, and it is seasoned with North American chili varieties instead of smoked paprika. The Mexican version also includes vinegar instead of wine. Chorizo is available in B.C. grocery stores, but the following recipe is included in case you can't find any. The recipe is essentially a general hybrid of Spanish and Mexican types. It follows the Mexican approach by using ground pork, with lots of heat coming from Cayenne pepper powder. But it also includes Spanish ingredients like paprika and wine. The end result is a versatile crumbled sausage useful in all sorts of dishes (see for example the Spanish-Italian pasta recipe on p. 67). If you are looking for a quick way to try it out, it makes an excellent topping for a slightly spicy pizza.

| | | |
|---|---|---|
| ½ lb | Ground pork, refrigerated (5°C) or frozen and just thawed | Add all of the ingredients to a bowl and mix them with a clean spoon or spatula (soak your utensils a few moments in clean hot water to be sure). Try to press, smear, and stretch out the meat during mixing, until everything is a bit tacky to the touch. Cover with cellophane, place in refrigerator (5°C), let the sausage rest for 1-2 hours before browning it. |
| 1-2 cloves | Garlic, minced | |
| ½ tsp | Salt (kosher, sea, or pickling) | |
| 1 tsp | Black pepper | |
| ½ tsp | White pepper | |
| 1½ tsp | Smoked paprika | |
| ⅛-¼ tsp | Cayenne | |
| 2 tbsp | Wine, white, dry (Riesling) | |
| 1 tbsp | Olive oil | |
| ¼ tsp | Baking soda | |

# Resources

In the following section, text coloured in light magenta signifies the location of a hyperlink in the electronic (.pdf) version of this book.

# Mushroom groups & info.[1]

**International:**
International Mycological Society (21)

**North America:**
North American Mycological Association (100)
*Mushroom* (Journal) (105)

**British Columbia:**
Interior B.C. Mushroom ID, Discussion, & Info.
Pacific Forestry Centre DAVFP herbarium collection
Vancouver Mycological society

**Europe:**
European Mycological Institute
European Mycological Association (9)
Journées européennes du Cortinaire (for *Cortinarius*)

**British Isles:**
British Isles Fungus Groups (28)
The Fungus Conservation Trust
Secret Sunday Mushroom Club
Irish Fungal Society
Northern Ireland Fungus Group

**Central Europe:**
Austrian Mycological Society
Mycology club of Brussels
German Mycology Society (20)
Society of Luxembourg Naturalists
Netherlands Mycological Society (12)
Swiss Fungi (National data and information centre)
Federation of Swiss Mushroom Associations (35)

**Western Europe:**
Mycology Society of France (49)
Iberian Mycological Society (272)

**South America:**
Carlos Spegazzini Mycological Association (Argentina)
Fungi Foundation (Chile)

**Africa:**
Fungi of Egypt
African Centre for Res. & Tech. Innov. (Univ. of Benin)
Mycological Society of Grand Canary
The Mycological Societies of Southern Africa

**Vancouver Island:**
South Vancouver Island Mycological Society
Vancouver Island Mushroom Identification and Info Group

**Nordic Countries**
Danish Mycological Society
Finnish Mycological Society
Icelandic Institute of Natural History (checklist)
Norwegian Association for Mycology & Foraging (41)
Local & regional mushhroom clubs in Sweden (Paper) (20)

**Eastern Europe:**
Czech Mycological Society (50)
Estonian Mycological Society
Hungarian Mycological Society
Latvian Mycological Society
Lithuanian Mycological Society

**Southern Europe:**
Mycological Society MycoBH (Bosnia & Herzegovina)
Cyprus Mycological Association
Croatian Mycological Society
Greek Mushroom Society
Italian Mycological Union (42)
Macedonian Mycological Society
Malta Mycological Association
Serbian Mycological Society
Slovenian Mycological Association (5)
Useful links (Slovenia)(8)

**Asia:**
Asian Mycological Association (10) China (Mainland, Taiwan), India, Indonesia, Japan, Korea, Malaysia, Pakistan, Philippines, Thailand)

**Australia / Oceana:**
Australasian Mycological Society
Fungi Group - Field Naturalists Club of Victoria
Australian Wild Mushroom Hunters
New Caledonia Mycological Society
New Zealand Mycological Society (2)

---

1    (The number of groups or member organizations listed at a site is shown in parentheses where applicable)

# Books, magazines, & websites[1]

Field and identification guides

This list includes modern references relevant to the British Columbia and the Pacific Northwest, along with some older ones. Guides published during the 20th century usually have outdated taxonomy and information (sometimes badly outdated) and should be used in conjunction with more modern references. They are listed here because a) you may run across used copies, b) it provides some awareness of mycologists who have helped to build our current knowledge, c) during identification, sometimes it is useful to compare (cautiously and carefully) images during identification to understand how a species concept has evolved (especially when others are still using older field guides to describe something to you), and d) there are always some who want to compare historical taxonomy, historical names, older descriptions, etc. to newer ones. In the Pacific Northwest the names, images, taxonomic information, and information concerning edibility are most current and reliable in the recent guides by Beug (2021), MacKinnon & Luther (2021), and Winkler (2022). If you live in the region, you should focus on those ones.

Arora, D. 1986. *Mushrooms demystified: A comprehensive guide to the fleshy fungi.* 2nd Ed. Ten Speed Press, Berkeley. 959pp. Reviewed at Goodreads.

Arora, D. 1991. *All That the Rain Promises and More: A Hip Pocket Guide to Western Mushrooms.* Ten Speed Press, Berkeley. 263pp. Reviewed at Goodreads.

Bandoni, R. and Szczawinski, S. 1964. *Guide to common mushrooms of British Columbia.* B.C. Provinicial Museum, Victoria, B.C. Handbook 24. 179pp.

**Beug, M. 2021. *Mushrooms of Cascadia: Illustrated key.* The Fungi Press, 314pp. Batavia, Ill.**

Davis, M., Sommer, R., and Menge, J.A. 2012. *Field guide to mushrooms of western North America.* Univ. of Calif. Press. 459pp.

Exeter, R.L., Norvell, L., and Cázares, E. 2006. *Ramaria of the Pacific Northwestern United States.* U.S. Dept. Int. BLM, Salem, Oregon. 157pp.

Groves, J. W. (addendum - Redhead, S.) 1979. *Edible and poisonous mushrooms of Canada.* Ag. Canada, Research Branch Publication 1112. Hull, Québec, 326pp.

Kroeger, P., Kendrick, B., Ceska, O., and Roberts, C. 2012. *The outer spores: Mushrooms of Haida Gwaii.* Mycologue Publications, Sidney, B.C. 189pp.

Lincoloff, G. (ed.) 1981. *Simon & Schuster's guide to mushrooms.* Simon and Schuster, New York. 511pp.

Lincoloff, G. 1981. *The Audubon Society field guide to North American mushrooms.* Knopf, NY. 926pp.

**MacKinnon, A. and Luther, K. 2021. *Mushrooms of British Columbia.* Royal B.C. Museum, Victoria, B.C., Canada. 498pp.**

McKenney, M. and Stuntz, D.E. (revised by Ammirati, J.F.). 1987. *The new savory wild mushroom.* D&M Publishers, Vancouver. 249pp.

Miller, O. 1972. *Mushrooms of North America.* E.P. Dutton & Co., N.Y. 360 pp.

Parker, H. 1996. *Alaska's mushrooms: A practical guide.* Alaska Northwest Books. Anchorage. 92pp.

Phillips, R. 1991. *Mushrooms of North America.* Firefly Books, Buffalo, NY, 319pp.

Sept, J.D. 2008. *Common Mushrooms of the Northwest.* Calypso Publishing, Sechelt, B.C. 94pp.

Schalkwyk, H. 1975. *Mushrooms of the Edmonton area: edible and poisonous.* H. Schalkwyk, 32pp.

Schalkwyk-Barendsen, H. 1991. *Mushrooms of northwest North America.* Lone Pine, Redmond, Wash. 414pp.

Smith, A. 1975. *A field guide to western mushrooms.* Univ. of Michigan Press, Ann Arbor, Mich. 280pp.

**Trudell, S. 2022. Mushrooms of the Pacific Northwest. Revised Edition. Timber Press. Portland, Oregon. 415pp.**

Trudell, S. and Ammirati, J. 2009. *Mushrooms of the Pacific Northwest.* Timber Press. Portland, OR, 347pp.

Underhill, J. 1979. *Guide to western mushrooms.* Hancock House, North Vancouver, B.C. 32pp.

Weber, N. *A morel hunter's companion.* 1988. Two Peninsula Press, Lansing, Mich. 208pp.

Winkler, D. 2011. *A field guide to edible mushrooms of the Pacific Northwest.* Harbour Publishing, China (foldout pocket guide).

**Winkler, D. 2022. Fruits of the forest: A field guide to Pacific Northwest edible mushrooms. The Mountaineers Books, Seattle, Wash. 384pp.**

---

1    Useful for those in the Pacific Northwest.

Books

Cook, L. 2013. The mushroom hunters: On the trail of secrets, eccentrics, and the American Dream. Ballantine Books, N.Y. 320pp. Reviewed at Goodreads.
Lincoff, G. 2017. *The complete mushroom hunter: An illustrated guide to foraging, harvesting, and enjoying wild mushrooms* (Revised). Quarto Publishing Group, Beverly, Massachusetts, U.S.A. 208pp. Reviewed at Goodreads.

Children's books

Boddy, L. 2021. Humongous fungus (underground and all around). DK Publishing, N.Y. 64pp. Synopsis at Strong Nations.
Gravel, E. 2018. The mushroom fan club. Drawn & Quarterly, Montréal, QC. 56pp. Reviewed at Goodreads.
Heller, R. 1992. Plants that never bloom. Grosset & Dunlap, N.Y. 48pp. Reviewed at Goodreads.

Magazines

*Fungi* https://www.fungimag.com/
*McIlvainea: Journal of American amateur mycology* (NAMA publication) https://namyco.org/mcilvainea.php
*The Mushroom* https://www.themushroom.pub/
*The Mushroom Grower's Newsletter* https://www.mushroomcompany.com/

Data, Software, & Education

Edible Wild Mushrooms Vancouver Island BC (information, courses) https://www.westcoastforager.com/
E-flora BC (collection data) https://ibis.geog.ubc.ca/biodiversity/eflora/fungi.html
Matsiman (picker information, focus on matsutake) https://www.matsiman.com/
Mushroom Observer https://mushroomobserver.org/
Gibson, I. 2020. MycoMatch (Matchmaker): Mushrooms of the Pacific Northwest. *Version 2.4*.
The Fungi of California https://www.mykoweb.com/CAF/
Tom Volk's Fungi https://botit.botany.wisc.edu/toms_fungi/
USDA Fungal databases - Saccardo (taxonomic information) https://nt.ars-grin.gov/fungaldatabases/Saccardo/Saccardo.cfm

# Glossary

**Ascomycete**—Fungi with spores that develop within an ascus. At the cellular level in fungi, ascii are microscopic membranous sac-like structures within which spores develop for sexual reproduction. In higher fungi they line fertile tissues on upper or outer cap surfaces. Ascomycetes include a variety of moulds, mildews, and yeasts, as well as higher fungi like morels and truffles.

**Basidiomycete**—Fungi with spores that develop in/on a basidium. At the cellular level in fungi, basidia are microscopic club-shaped structures that bear spores developed for sexual reproduction. In higher fungi they line fertile tissues under caps, on the surface of gills, pores, teeth, etc. Many of the large fruiting bodies people know as mushrooms are basidiomycetes.

***Béchamel* sauce**—(Pronounced *Bay-shah-mel*.) White sauce used in French cooking. To make it, heat 1¼ cups milk in a sauce pan (med. heat) until it just begins to bubble at the edges then remove from heat. Melting 2 tbsp butter on (med. heat) and add 1½-2 tbsp flour and make a roux, stirring for 1 min. Add the hot milk until the broth thickens, and then seasoning to taste with ground salt, pepper, and perhaps ground nutmeg.[1]

**Bioaccumulation**—Over time, the gradual accumulation and concentration of a substance (usually a trace element, pollutant, or contaminant) with an organism.

**Bioactive**—Having any sort of biological effect.

**Caramelization**—During cooking, a chemical reaction resulting in the browning of sugars and the development of a sweet, nutty flavour. A brown colour develops from the formation of certain sugar polymers (caramelans, caramelens, and caramelins). The caramel flavour results from the release of volatile chemicals (e.g. diacetyl).

***Carpaccio***—Typically an appetizer consisting of beef, venison, veal, salmon, or tuna sliced or pounded thin. It was invented in Venice, Italy in 1963 by Giuseppe Cipriani. Currently, the term is also applied to thinly sliced truffles soaked in brine and then conserved in olive oil.

**Deglaze**—Adding liquid to a pan to release the fond for easier scraping and dispersion (see *fond*).

**Family**—In taxonomy, a rank in the classification of organisms that is between genus and order, i.e. there can be several genera in a family, and several families in an order.

**Fond**—French for 'base,' referring to the brown residue that forms and sticks to the bottom of a skillet or pan after a sauté or roast. It often contains a good concentration of flavours that can be restored to the liquids in a dish through scraping.

**Genus**—In taxonomy, a rank in the classification of organisms that is between species and family, i.e. there can be several species in a genus, and several genera in a family.

**Green onion**—An older scallion.

**Hollandaise sauce**—French for 'Dutch sauce.' The sauce is composed of egg yolk and melted butter, along with either lemon juice, white wine, or vinegar.

***Herbs de Provence***—A blend of herbs widely grown and popular in southeastern France. Basic ingredients usually include fennel, marjoram, parsley, rosemary, tarragon, and thyme. The wide variety of recipes for this blend might also include basil, bay leaves, savory, chervil, sage, oregano, mint, and/or lavender. The blend is used to flavour roasted chicken, omelettes, tomato sauce, pasta dishes, etc.

**Immunomodulation**—A change in the body's immune system due to the action of substances that stimulate or inhibit it.

**Mirin**—A sweet cooking wine from Japan, made from fermented rice.

**Mycelium**—The vegetative (non-sexual) colonial form of fungi consisting of a network or matrix of branching filamentous cells or *hyphae* that grow and spread throughout a substrate or living host (see also *hyphae*).

**Mornay sauce**—A sauce made by adding shredded cheese and melting it in a Béchamel sauce (q.v. above). Cheeses added include Gruyère, Emmenthal, Parmesan, or mild white cheddar.

**Mycorrhizae**—Literally, 'fungus-root.' An association, normally symbiotic, between soil fungi and plant roots. It is a key relationship that allows plants and trees to thrive in may different ecosystems. In the symbiosis, energy-containing compounds from the plant are exchanged for mineral nutrients gathered by the fungal mycelium. The kind of symbiosis commonly found in basidiomycete mushrooms is 'ectomycorrhizal' (in which the mycelium forms a sheath-like

---

1    Cunningham, M. 2004; Soulard, J. 2022.

structure around small and abbreviated or club-shaped plant roots). Not all mycorrhizal fungi are ectomycorrhizal, but 90% of plant species form some type of mycorrhizal relationship with soil fungi.

**Roux**—A combination of fat and flour that forms a paste that can become a thickening agent in soups, etc.

**Scallion**—A young green onion.

**Shallot**—A variety of onion widely used in cooking. Shallots grow somewhat like garlic, with clustered bulbs and heads with cloves. Colour of the bulbs varies from brown to red and the flesh is off-white with green or magenta hues. In Québec, anglophones may refer to scallions as shallots, while calling actual shallots 'French shallots.'

**Species**—In taxonomy, a rank in the classification of organisms that is under genus, i.e. there can be several species in a genus (and there can be several varieties, ecotypes, strains, etc. within a species). Species are theoretically distinct in terms of their ability to mate and produce fertile offspring. In reality there are many exceptions and complications for this distinctiveness (e.g. hybridization). 'Species' is therefore a convenient and useful concept for organizing our taxonomic knowledge about different organisms, but there is no reason that biology in the real world would slavishly adhere to the concept.

**Spring onion**—A type of young onion with a small, rounded bulb at the base, planted for springtime harvest. Slightly more mature than green onions or scallions and with a stronger flavor, they are nevertheless still younger than regular onions.

**ssp.**—Species (plural).

**Stem**—(Also stipe). In mushrooms, the portion of the fruiting body that attaches the pileus (cap) to the substrate (soil, log, woody stem, etc.) Stems can be vestigial or lacking in some species. In other species, they can bear various key diagnostic features (membrane remnants, ridges, glands, bruises and staining) that help to characterize a species.

**Taxonomy**—The study of classifying things.[1]

**Trace element**—A chemical element that only exists in minute amounts in a particular organism or environmental sample.

**Umami**—One of the basic tastes along with sweet, sour, bitter, and salty. From Japanese, 'savouriness.' A characteristic of broths and cooked meats, people experience umami through taste receptors stimulated by glutamates and nucleotides. These taste receptors are unique, rather than a combination of other types of receptors, so it is a basic taste. Umami foods include shellfish and fish; tomatoes; mushrooms; extracts of meat, hydrolyzed vegetable products or yeasts; cheeses; various sauces, kelp and other algae, etc.

**Zest**—The outermost skin or epidermal layer of citrus fruits. The zest contains intense flavours useful in cooking. It is scraped from the rind to provide flavour without involving excessive bitterness of the rind or acidity of the fruit juices.

---

1    The study of people who classify things has been called the Psychology of Genre. See Vanderbilt, T. 2016.

# References cited

These are more complete references corresponding to the footnotes in each chapter. Some of these references also appear in the bibliographies for field guides and recipe books. Text appearing in light magenta indicates the existence of a hyperlink to the source (the .pdf version of this book has the clickable link).

Acton, B. and Duncan, P. 1992. Making wines like those you buy. G.W. Kent Inc., 160pp. Reviewed at *Goodreads*.

Adams, D. 1978. The hitchikers guide to the galaxy. (Radio series). *British Broadcasting Company, Radio 4*.

Aisala, H., Manninen, H., Laaksonen, T., Linderborg, K., Myoda, T., Hopia, A., and Sandell, M. 2020. Linking volatile and non-volatile compounds to sensory profiles and consumer liking of wild edible Nordic mushrooms. *Food Chem.* 304:125403.

Allen, K. and Bennett, J. 2021. Tour of truffles: aromas, aphrodisiacs, adaptogens, and more. *Mycobiol.* 49:201-212.

Allen, L. 2020. Roasted red pepper pasta. *Tastes Better from Scratch*.

Ami. 2005. Finnish baked mushrooms recipe. *Food.*.

Arora, D. 1986. Mushrooms demystified: A comprehensive guide to the fleshy fungi. 2nd Ed. Ten Speed Press, Berkeley. 959pp. Reviewed at *Goodreads*.

Arpin, N. and Fiasson, J. 1971. The pigments of basidiomycetes: their chemotaxonomic interest. Pp. 63-98 in: Peterson, R. Evolution in the higher Basidiomycetes. U. Tenn. Press, Knoxville, Tenn.

Awada, R. Easy salmon Florentine. *Healthy Fitness Meals*.


Barros, L., Cruz, T., Baptista, P., Estevinho, L., and Ferreira, I. 2008. Wild and commercial mushrooms as source of nutrients and nutraceuticals. *Food & Chem. Tox.* 46:2742-2747.

Battison, G., Degetto, S., Gerbasi, R., and Sbrignadello, G. 1989. Radioactivity in mushrooms in northeast Italy following the Chernobyl accident. *J. Env. Radioactivity* 9:53-60.

Beeler, J. 2000. Potato casserole. *allrecipes*.

Bellesia, F., Pinetti, A., Bianchi, A. and Tirillini, B. 1998. The volatile organic compounds of black truffle (*Tuber melanosporum* Vitt.) from Middle Italy. *Flavour Fragr. J.*, 13:56-58.

Bellesia, F., Pinetti, A., Tirillini, B., and Bianchi, A. 2001. Temperature-dependent evolution of volatile organic compounds in *Tuber borchii* from Italy. *Flav. & Fragr. J.* 16:1-6.

Bellesia, F., Pinetti, A., Tirillini, B., Paolocci, F., Rubini, A., Arcioni, S., and Bianchi, A. 2002. The headspace volatiles of the Asian truffle *Tuber indicum* Cooke et Mass, *J. Essent. Oil Res.* 14:3-5.

Bell, G. 2021. Roasted pork tenderloin with wild mushrooms – recipe! *Live, Love, Laugh, Food*.

Beluhan, S. and Ranogajec, A. 2011. Chemical composition and non-volatile components of Croatian wild edible mushrooms. *Food Chem.* 124:1076-1082.

Benjamin, D. 1995. Mushrooms: poisons and panaceas. A handbook for naturalists, mycologists, and physicians. W. H. Freeman & Co., N.Y. 422pp. Reviewed at *Goodreads*.

Bergo, A. 2014c. Mangalista pork chops with chanterelle-skyr sauce. *Forager Chef*.

Bernheimer, A. and Oppenheim, J. 1987. Some properties of flammutoxin from the edible mushroom *Flammulina velutipes*. *Toxicon* 25: 1145-1152.

Beug, M. 2020. Summary of the poisoning reports in the NAMA Mushroom Poisoning Case Registry: 2018 through 2020. *Namykco.org*.

Bittman, M. 2018. Pasta fritatta. *New York Times (Cooking)*.

Bobeck, M. 2024. Creamed carrots, recipe from the *Titanic* (ship - not the movie). *Food.*.

Bonito, G., Trappe, J, Donovan, S. and Vilgalys, R. 2011. The Asian black truffle *Tuber indicum* can form ectomycorrhizas with North American host plants and complete its life cycle in non-native soils. *Fungal Ecol.* 4:83-93.

Boruah, A. 2023. *Craterellus cornucopioides* (L.)/*Craterellus odoratus* (Schwein.) Fr. (Black Chanterelle). Pp. 142-164 in: Sharma, A., Bhardwaj, G., and Nayik, G. (eds.) Phytochemistry and nutritional composition of significant wild medicinal and edible mushrooms: traditional uses and pharmacology. Royal Soc. of Chemistry, London. 438pp.

Brian. 2015. Sautéed Enoki Mushrooms. *Food52*.

Brunnert, H., and Zadražil, F. 1983. The translocation of mercury and cadmium into the fruiting bodies of six higher fungi. *Eur. J. Appl. Microbiol. Biotechnol.* 17:358–364.

Bruxel, A. 2021. Chicken roulade - with shiitake mushrooms, sage and prosciutto. *Home Chef Seattle*.

Bull, N. 2024. Easy canned tuna pasta (ready in 15 minutes!). *Salt & Lavender*.

Butler, A. 2022. Swedish meat patties & chanterelle sauce recipe. *Just Like Granny*.

Butts, V. 2024. Finnish baked mushrooms. *Just A Pinch Recipes*.

Buzzini, P., Gasparetti, C., Turchetti, B., Cramarossa, M., Vaughan-Martini, A., Martini, A., Pagnoni, U., and Forti, L. 2005. Production of volatile organic compounds (VOCs) by yeasts isolated from the ascocarps of black (*Tuber melanosporum* Vitt.) and white (*Tuber magnatum* Pico) truffles. *Arch. Microbiol.* 184:187–193.

Cai, H., Liu, X., Chen, Z., Liao, S., and Zou, Y. 2013. Isolation, purification and identification of nine chemical compounds from *Flammulina velutipes* fruiting bodies. *Food Chem.* 141:2873-2879.

Cascina Alberta. 2023. Truffle history. *Cascina Alberta*.

Cebulah, T., and Wicks, L. 2016. What do I substitute for mirin? *Cooking Light*.

Chen, J., Li, J. M., Tang, Y., Xing, Y., Qiao, P., Li, Y., Liu, P., and Guo, S. 2019. Chinese black truffle-associated bacterial communities of *Tuber indicum* from different geographical regions with nitrogen fixing bioactivity. *Front. Microbiol.* 10:2515.

Chen, J., Murat, C., Oviatt, P., Wang, Y., and Le Tacon, F. 2016. The black truffles *Tuber melanosporum* and *Tuber indicum*. Pp. 19-32 in Zambonelli, A., Iotti, M., and Murat, C. (eds.) *True Truffle (Tuber spp.) in the World: Soil Ecol., Syst. & Biochem*. Spring Cham., Switzerland. 436pp.

Chen, X., Huynh, N., Cui, H., Zhou, P., Zhang, X., and Yang, B. 2018a. Correlating supercritical fluid extraction parameters with volatile compounds from Finnish wild mushrooms (*Craterellus tubaeformis*) and yield prediction by partial least squares regression analysis. *RSC Adv.* 8:5233-5242.

Chen, X., Yu, J., Cui, H., Xia, S., Zhang, X., and Yang, B. 2018b. Effect of temperature on flavor compounds and sensory characteristics of Maillard reaction products derived from mushroom hydrolysate. *Molecules* 23:247.

Chicken Farmers of Canada. 2024. Chicken ballotine with mushroom tarragon and chèvre served with Israeli couscous and apricot salad. *Chicken.ca*.

Chu, P., Sun, H., Ko, J., Ku, M., Lin, L. Lee, Y., Liao, P., Pan, H., Lu, H., and Lue, K. 2017. Oral fungal immunomodulatory protein-*Flammulina velutipes* has influence on pulmonary inflammatory process and potential treatment for allergic airway disease: A mouse model. *J. Microbiol., Immun. & Inf.* 50:297-306.

Chung, H., Kim, D., and Lee, S. 2002. Mycorrhizal formations and seedling growth of *Pinus desiflora* by *in vitro* synthesis with the inoculation of ectomycorrhizal fungi. *Mycology* 30:130-175.

Colpaert, H. and Van Laere, A. 1996. A comparison of the extracellular enzyme activities of two ectomycorrhizal and a leaf-saprotrophic basidiomycete colonizing beech leaf litter. *New Phytol.* 133:133-141.

Colpaert, J.V., Wevers, J.H., Krznaric, E., and Adriensen, K. 2011. How metal-tolerant ecotypes of ectomycorrhizal fungi protect plants from heavy metal pollution. *Ann. For. Sci.* 68:17–24.

Costantin, J. and Matruchot, L. 1898. Botanique agricole - Essai de culture du "*Tricholoma nudum.*" *Revue de Mycol.* Supp. 7:63-67.

Culleré, L., Ferreira, V., Chevret, B., Venturini, M., Sánchez-Gimeno, A., and Blanco, D. 2010. Characterisation of aroma active compounds in black truffles (*Tuber melanosporum*) and summer truffles (*Tuber aestivum*) by gas chromatography–olfactometry. *Food Chem.* 122:300-306.

Culleré, L., Ferreira, V., Venturini, M. E., Marco, P., and Blanco, D. 2013. Chemical and sensory effects of the freezing process on the aroma profile of black truffles (*Tuber melanosporum*). *Food chem.* 136:518-525.

Cunningham, M. 2004. White sauce or bechamel sauce. *Epicurious*.


Dairy Farmers of Canada. 2021. Canadian cheddar cheese soup. *Dairy Farmers of Canada*.

Danai, O., Ezov, N., Levanon, D., and Masaphy, S. 2008. Introduction of new exotic mushroom species into cultivation in Israel. *Israel J. Plant Sci.* 56:295-301.

Daniel. 2022. Truffle chicken roulade/chicken ballotine, truffle mousse, preserved lemon gastrique. *Daniel's Gongbang*.

Dash, T. 2020. Salisbury steak – made with beef, bison, chicken, or turkey! *Boulder Lacavore*.

De, J., Nandi, S., and Acharya, K. 2022. A review on Blewit mushrooms (*Lepista* sp.) transition from farm to pharm. *J. Food Proc. & Preserv.* 00:e17028.

Deo, G., Khatra, J., Bukttar, S., Li, W., Tackaberry, L., Masicotte, H., Egger, K., Reimer, K., and Lee, C. 2019. Antiproliferative, immunostimulatory, and anti-inflammatory activities of extracts derived from mushrooms collected in Haida Gwaii, British Columbia (Canada). *Int. J. Med. Mush.* 21:629-643.

Drennan, F. 2010. Wild winter soups. *Country Kitchen* Jan. 2010, p. 73.

Drummond, R. 2013. Quick and easy roasted red pepper pasta. *Food Network*.

Du, J., Guo, H., Li, Q., Forsythe, A., Chen, X., and Yu, X. 2018. Genetic diversity of *Lepista nuda* (Agaricales, Basidiomycota) in Northeast China as indicated by SRAP and ISSR markers. *PLoS ONE* 13:e0202761.

Duckhorn Wine Co. 2024. Grilled pork tenderloin with white truffle oil, shiitake mushrooms and blue cheese sauce. *Goldeneye Winery*.


Eataly. 2015. A guide to Italian olive varieties. *Eataly*.


Faisson, J. and Arpin, N. 1967. Recherches Chimotaxinomiques sur les champignons. V. Sur les carotenoides mineurs de *Cantharellus tubaeformis*. *Bull.Soc. Chim. Biol.* 49:537-542.

Fang, D., Yang, W., Kimatu, B., Zhao, L., An, X., and Hu, Q. 2017. Comparison of flavour qualities of mushrooms (*Flammulina velutipes*) packed with different packaging materials. *Food Chem.* 232:1-9.

Feng, B., Wang, X., Ratkowsky, D.,Gates, G., Lee, S., Grebenc, T., and Yang, Z. 2016. Multilocus phylogenetic analyses reveal unexpected abundant diversity and significant disjunct distribution pattern of the Hedgehog Mushrooms (*Hydnum* L.). *Sci. Rep.* 6:25586.

First Nature. 2021. *Pholiota adiposa* (Batsch) P. Kumm. *First Nature*.

First Nature. 2022. *Lepista nuda* (Bull.) Cooke - Wood blewit. *First Nature*.

Fischer, D. and Bessette, A. 1992. Edible wild mushrooms of North America: A field to kitchen guide. *Univ. Of Texas Press*, Austin, Texas, U.S.A. 254pp. Reviewed at *Goodreads*.

Fisher, T. 2024. Zinfandel black trumpet steak sauce. *Restauranteur*.

Fons, F., Rapior, S., Eyssartier, G., Bessière, J.-M. 2003. Volatile compounds in the *Cantharellus*, *Craterellus* and *Hydnum* genera. *Cryptogamie Mycologie*. 24. 367-376.

Force of Nature. 2020. Bison Salisbury steaks with mushroom gravy. *Force of Nature*.

Foster, K. 2015. What's the difference between regular olive oil and extra-virgin olive oil? *Kitchn*.

Four Magazine. 2014. Recipe: Truffled ballotine of chicken. *Four Magazine*.

Fukushima, M., Ohashi, T., Fujiwara, Y., Sonoyama K., and Nakano, M. 2001. Cholesterol-lowering effects of maitake (*Grifola frondosa*) fiber, shiitake (*Lentinula edodes*) fiber, and enokitake (*Flammulina velutipes*) fiber in rats. *Exp. Biol. & Med.* 226:758-765.

Fuller, J. 2017. Seven ways to tell if your olive oil is fake. *Epicurious*.

Gaitan-Heernandez, R. and Baez Rodriguez, R. 2008. Mycelial growth of *Lepista nuda* native wild strains on culture media with different organic supplements. *Rev. Mex. Mic.* 26:41-49.

García-Montero, L., Quintana, A., Valverde-Asenjo, I., and Díaz, P. 2009. Calcareous amendments in truffle culture: a soil nutrition hypothesis. *Soil Biol. & Biochem.* 41:1227-1232.

Gibson, I. 2020. Annotated checklist for larger fungi on Vancouver Island. *South Vancouver Island Mycological Society*.

Gone71°N. 2022. Hedgehog mushroom | *Hydnum repandum* (recipe). *Gone71°N*.

Gordon Ramsey Restaurants. 2022. The history of Beef Wellington is well worth celebrating. *Gordon Ramsey Restaurants* 28 Sept. 2022.

Govorushko, S., Rezaee, R., Dumanov, J., Tsatsakis, A. 2019. Poisoning associated with the use of mushrooms: A review of the global pattern and main characteristics. *Food & Chem. Toxicol* 128:267-279.

Grace, L. 2022. Truffle mashed potatoes. *Chef not required... Recipes from a home cook*.

Grigson, J. 1975. The mushroom feast. Alfred A. Knopf, New York. 305pp. Reviewed at *Amazon*.

Guo, H., Diao, Q., Hou, D., Li, Z., Zhou, Z., Feng, T., and Liu, J. 2017. Sesquiterpenoids from cultures of the edible mushroom *Craterellus cornucopioides*. *Phytochem. Lett.* 21:114-117.

Guo-ying, Z, Liang, G., Lin, L., and He, L. 2010. rDNA internal transcribed spacer sequence analysis of *Craterellus tubaeformis* from North America and Europe. *Can. J. Microbiol.* 57:29-32.

Hagenbo, A., Kyaschenko, J., Clemmensen, K., Lindahl, B., and Fransson, P. 2018. Fungal community shifts underpin declining mycelial production and turnover across a *Pinus sylvestris* chronosequence. *J. Ecol.* 106:490– 501.

He, Z., Chen, Z., Bau, T., Wang, G., and Yang, Z. 2023. Systematic arrangement within the family Clitocybaceae (Tricholomatineae, Agaricales): phylogenetic and phylogenomic evidence, morphological data and muscarine-producing innovation. *Fungal Diversity* 123:1–47.

Helena. 2021. Funnel chanterelle soup. *Nordic Forest Foods*.

Herrero de Aza, C., Armenteros, S., McDermott, J., Mauceri, S., Olaizola, J., Hernández-Rodríguez, M., and Mediavilla, O. 2022. Fungal and bacterial communities in *Tuber melanosporum* plantations from Northern Spain. *Forests* 13:385.

Iadanza, N. 2015. Chanterelles 101 - post script. *Mad About Mushrooms* Feb. 2015.

Jabeen, S., Ilyas, S., Niazi, A., and Kkalid, A. 2012. Diversity of ectomycorrhizae associated with *Populus* spp. growing in two different ecological zones of Pakistan. *Int. J. Ag. & Biol.* 14:681-688.

Jacobson, A. 2016. Nametake: Enoki mushrooms in mirin and soy sauce. *One Green Planet*.

Jawad, Y. 2022. Creamy roasted red pepper pasta. *Feel Good Foodie*.

Jedidi, I., Ayoub, I., Philippe, T., and Bouzouita, N. 2017. Chemical composition and nutritional value of three Tunisian wild edible mushrooms. *Food Meas.* 11:2069–2075.

Jeon, J., Lee, K., Lee, K., Kim, M., Kim, I., and Kim, Y. 2020. Characteristics and pedigree selection of a shortened cultivation period strain in *Lepista nuda*. *J. Mush.* 18:331–338.

Jiang, Q., Zhang M, and Mujumdar A. 2020. UV induced conversion during drying of ergosterol to vitamin D in various mushrooms: Effect of different drying conditions. *Trends Food Sci. Technol.* 105:200-210.

Jiang, X., Chu, Q., Li, L., Qin, L., Hao, J., Kou, L., Lin, F., and Wang, D. 2018. The anti-fatigue activities of *Tuber melanosporum* in a mouse model. *Exp. & Ther. Med.* 15: 3066-3073.

Jolly, N. 2020a. Olives: How does it grow? Part 1. Olive oil: how is it made? *True Food TV*.

Jolly, N. 2020b. Olives: How does it grow? Part 2. Olive oil: how is it made? *True Food TV*.

Jordan, P. and Wheeler, S. 1995. The ultimate mushroom book. Anness Publishing Lt., London. 246pp. Reviewed at *Goodreads*.

Judy. 2017. Enoki mushrooms with garlic and scallion sauce. *The woks of life: A culinary genealogy*.

Juvenalis, Decimus Junius. ca. 100-127 A.D. Saturae (*Book V*).

Kalač P. 2001. A review of edible mushroom radioactivity. *Food Chem.* 75:29–35.

Kang, Z., Li, X., Li, Y., Ye, L., Zhang, B., Zhang, X., Penttinen, P. and Gu, Y. 2022. Black truffles affect *Quercus aliena* physiology and root-Associated nirK- and nirS-type denitrifying bacterial communities in the initial stage of inoculation. *Front. Microbiol.* 13:792568.

Karadsheh, S. 2023. Roasted red pepper pasta. *The Mediterranean Dish*.

Kasuya, M. 1996. *In vitro* ectomycorrhizal formation in *Picea glehnii* seedlings. *Mycorrhizae* 6:451–454.

Kiss, M., Csóka, M., Győrfi, J., and Korány, K. 2011. Comparison of the fragrance constituents of *Tuber aestivum* and *Tuber brumale* gathered in Hungary. *J. Appl. Bot. Food Qual.* 84:102-110.

Kosanić M., Ranković B., Stanojković T., et al. 2019. *Craterellus cornucopioides* edible Mushroom as source of biologically active compounds. *Nat. Prod. Comms. 14(5):1-6*.

Kostiainen, E. 2007. 137Cs in Finnish wild berries, mushrooms and game meat in 2000–2005. *Bor. Env. Res.* 12:23–28.

Kuo, M. 2020. *Hydnum repandum*. *MushroomExpert*.

Ladies Aid Society. ca. 1915. Cornflower Cook Book. Ladies Aid Society, 1st Presbyterian Church, Gary, Indiana, USA.

Laka. 2024. Pork tenderloin with oyster mushrooms and parsley. *Food.*.

Lee, S., Choi, K., and Oh, C. 1996. Sawdust cultures of *Lepista nuda*. *Kor. J. Mycol.* 24:274-279.

Li, H., Yang, W., Qu, S., Pei, F., Luo, X., Mariga, A., and Ma, L. 2018. Variation of volatile terpenes in the edible fungi mycelia *Flammulina velutipes* and communications in fungus-mite interactions. *Food Res. Int.* 103:150-155.

Li, J., Liang, H., Qiao, P., Su, K., Liu, P., Guo, S., and Chen, J. 2019. Chemical composition and antioxidant activity of *Tuber indicum* from different geographical regions of China. *Chem. & Biodiv.* 16:e1800609.

Lin, J., Wu, H., and Shi, G. 1975. Toxicity of the cardiotoxic protein, flammutoxin, isolated from the edible mushroom *Flammunlina velutipes*, *Toxicon* 13: 323-332.

Liu, J. 2007. Secondary metabolites from higher fungi in China and their biological activity. *Drug Discov. Ther.* 1:94-103.

Lu, H. 2018. The illegal ramen vendors of postwar Tokyo. *Atlas Obscura*.

Luard, E. 2004. Classic French Cooking. Recipes for mastering the French kitchen. Octopus Publishing Group, N.Y. 336pp.

Lydon, K. 2014. Truffle mushroom eggs Benedict. *Kara Lydon Nutrition*.

MacKinnon, A. and Luther, K. 2021. *Mushrooms of British Columbia*. Royal B.C. Museum, Victoria, B.C., Canada. 498pp.

Maclean, F. 2012. Mushroom stuffed pork tenderloin with truffled mushroom sauce. *London Unattached*.

Mallett, E. 2024. Hangar steak with black trumpet cream sauce. *The Recipe Circus*.

Manninen, H., Rotola-Pukkila, M., Aisala, H., Hopia, A., and Laaksonen, T. 2018. Free amino acids and 5'-nucleotides in Finnish forest mushrooms. *Food Chem.* 247:23-28.

Masood, H. 2022. Investigating the fluxes of radio-cesium ($^{137}$Cs) in Sweden. MS Thesis. Dept. of Earth Sci., Uppsala Univ., Uppsala, Sweden.

McEwan, M. 2021. This mushroom tart is the best of all savoury tarts—trust me. *Toronto Star (Food & Drink)* Dec. 18, 2021

Mello, A., Lumini, E., Napoli, C., Bianciotto, V., and Bonfante, P. 2015. Arbuscular mycorrhizal fungal diversity in the *Tuber melanosporum* brûlé. *Fungal Biol.* 119:518-527.

Miller, M. 2023. The devilish history of devilled eggs. *Tasting History with Max Miller*.

Montagné, P. 1938. *Larousse gastronomique* (1961 translation by N. Froud). Hamlyn Publishing Group Ltd., London. 1098pp. Reviewed at *Amazon*.

Moynier, MM. 1836. De la truffe, de lat traité complet de ce tubercle. Barba, Libraire , Palais-Royal ; galerie de Chartres. Legrand et Bergougnioyx, success. de V° C.Béchet, Quai des Grands-Augustins, n. 59. Available at U.S. Library of Congress.

Mushroom-Collecting.com. 2022. The "small chanterelles" (*Craterellus tubaeformis, Craterellus ignicolor*). *Mushroom-Collecting.com*.

Mustafa, A., Angelone, S., Nzekoue, F., Abouelenein, D., Sagratini, G., Caprioli, G., and Torregiani, E. 2020. An overview on truffle aroma and main volatile compounds. *Molecules* 25:598.

Mustonen, A., Määttänen, M., Kärjä, V., Puuka, K., Aho, J., Saarela, S., and Nieminen, P. 2018. Myo- and cardiotoxic effects of the wild winter mushroom (*Flammulina velutipes*) on mice. *Exp. Biol. & Med.* 243:639-644.

Nakamura, K. 1981. Mushroom cultivation in Japan. Asaki Publication House, Japan.

Napoli, C., Mello, A., Borra, A., Vizzini, A., Sourzat, P. and Bonfante, P. 2010. *Tuber melanosporum*, when dominant, affects fungal dynamics in truffle grounds. *New Phytol.* 185:237-247.

Niskanen, T., Liimatainen, K., Nuytinck, J., Kirk, P., Ibarguren, I., Garibay-Orijel, R., Norvell, L., Huhtinen, S., Kytövuori, I., Ruotsalainen, J., Niemelä, T., Ammirati, J., and Tedersoo, L. 2018. Identifying and naming the currently known diversity of the genus *Hydnum*, with an emphasis on European and North American taxa. *Mycologia* 110:890-918.

Novakovic, S. 2021. The potential of the application of *Boletus edulis, Cantharellus cibarius* and *Craterellus cornucopioides* in frankfurters: a review. *Earth & Env. Sci.* 854:012068.

O'Callaghan, Y., O'Brien, N., Kenny, O., Harrington, T., Brunton, N., and Smyth. T. 2015. Anti-inflammatory effects of wild Irish mushroom extracts in RAW264.7 mouse macrophage cells. *J. Med. Food.* 18:202-207.

Olive oils from Spain. 2022. *Olive Oils from Spain.*

O'Reilly, P. 2022. *Cantharellus tubaeformis* Fr. - Trumpet Chanterelle. *First Nature.*

Özdal, M. 2018. Determination of carbon and nitrogen sources for the production of mycelial biomass and exopolysaccharide by *Lepista nuda* in liquid culture. *Turk. J. Ag. -Food Sci. Tech.* 6:581-585.

Ozuna-Valencia, K., Moreno-Robles, A., Rodríguez-Félix, F., Moreno-Vásquez, M., Berraras-Urbina, C., Madera-Santana, T., Ruíz-Cruz, S., Bernal-Mercado, A., Armenta-Villegas, L., and Hernández, J. 2023. Black trumpet (*Craterellus cornucopiodes*). Chapter 9 in: Pandita, D. and Pandita, A. (eds.) Mushrooms: nutraceuticals and functional foods. CRC Press. Boca Raton, Fla. 386pp.


Pang, Z. 1993. Secondary fungal metabolites isolated from fruit bodies. PhD. Thesis, Dept. of Organic Chemistry 2, Lund Univ., Sweden.

Parisi, G. 2014. Pork tenderloin with wild mushrooms, ginger and scallions. *Food & Wine.*

Paulina. 2012. On the different varieties of Greek olives. *Isle of Olive.*

Pegler, D.N. 2003. Useful fungi of the world: the Shii-take, Shimeji, Enoki-take, and Nameko mushrooms. *Mycologist* 17(1):3-5.

Pépin, J. 2023. Duck à l'Orange. *Food & Wine.*

Piasecki, B. 2008. How To Make Cheddar Cheese Soup from Disney's EPCOT. *The Disney Blog.*

Pilz, D., Norvell, L., Danell, E., and Molina, R. 2003. Ecology and management of commercially harvested chanterelle mushrooms. Gen. Tech. Rep. PNW-GTR-576. Portland, OR: U.S. Department of Agriculture, Forest Service, Pacific Northwest Research Station. 83 pp.

Plattner, I., and Hall, I. 1995. Parasitism of non-host plants by the mycorrhizal fungus *Tuber melanosporum. Mycol. Res.* 99: 1367-1370.

Pliny the Elder. 77 AD. *Naturalis historia.* Translated in Bostock, J. 1855. *The Natural History. Pliny the Elder.* Taylor & Francis, Red Lion Ct., Fleet St., London. 1855.

Powell, G. 2022. Pittsburg potatoes recipe—Old cookbook show. *Glen And Friends Cooking.*

Powell, M. 2014. Medicinal Mushrooms - A Clinical Guide by Martin Powell. Mycology Press.


Rafuse, C. 1990. Canadian-Bred Tomatoes. Heritage Seed Program, *Seeds of Diversity* August 1990.

Reder, H. 2020. Chanterelle Gruyère tart. *Kitchen Stories.*

Richa. 2023. Bacon mushroom chicken roulade. *My Food Story.*

Rodrigues, A. 2008. Truffles. *Trufamania.*

Roques, J. 1832. *Histoire des champignons comestible et vénéneux, ornée de figures coloriées représentant les principales espèces dans leurs dimensions naturelles, où l'on expose leurs caractères distinctifs, leurs propriétés alimentaires et économique, leurs effets nuisible et les moyens deo s'en garantie ou d'y remédier, ouvrage utile aux amateurs des champignons, aux médecin, aux naturalístes,aux propriétaires ruraux, aux maires de villes, et de campagnes, etc.* Hocquart Aîně, Éditeur. Imprimerie de Casmir, Rue de la Vielle-Monnaie, près la rue des Lombards et la place du Châtelet. Paris.

Roques, J. 1841. *Histoire des champignons comestible et vénéneux (Deuxième édition).* Fortin, Masson, et Cie, Libraires-Éditeurs, 1 Place de l'Ecole-de-Médecine. Paris.

Rubini, A., Paolocci, F., Riccioni, C., Vendramin, G., and Arcioni, S. 2005. Genetic and phylogeographic structures of the symbiotic fungus Tuber magnatum. *Appl. & Env. Microbiol.* 71:6584-6589.


S., Lucille (threeovens). 2024. Roasted pork loin with wild mushroons, garlic, and sage pan *jus. Food.*.

Safina, R. and Sutton, J. 2003. Truffles: Ultimate luxury, everyday pleasure. John Wiley & Sons. 246p. Reviewed at *Goodreads*.

Saltarelli, R., Ceccaroli, P., Cesari, P., Barbieri, E., and Stocchi, V. 2008. Effect of storage on biochemical and microbiological parameters of edible truffle species. *Food Chem.,* 109:8-16.

Savini, S., Loizzo, M. R., Tundis, R., Mozzon, M., Foligni, R., Longo, E., Morozova, K., Scampicchio, M., Martin-Vertedor, D., and Boselli, E. 2017. Fresh refrigerated *Tuber melanosporum* truffle: Effect of the storage conditions on the antioxidant profile, antioxidant activity and volatile profile. *Eur. Food Res. & Tech.,* 243:2255-2263.

Schneider-Maunoury, L., Deveau, A., Moreno, M., Todesco, F., Belmondo, S., Murat, C., Courty, P., Jąkalski, M. and Selosse, M. 2020. Two ectomycorrhizal truffles, *Tuber melanosporum* and *T. aestivum,* endophytically colonise roots of non-ectomycorrhizal plants in natural environments. *New Phytol.* 225:2542-2556.

Shah, N., Marathe, S., Croce, D., Ciardi, M., Longo, V., Juilus, A., and Shamekh, S. 2022. An investigation of the antioxidant potential and bioaccumulated minerals in *Tuber borchii* and *Tuber maculatum* mycelia obtained by submerged fermentation. *Arch. Microbiol.* 204:64.

Shaw, H. 2010. Venison Swedish meatballs. *Hunter Gather Cook.*

Shaw, H. 2017. Duck a L'Orange. *Hunt Gather Cook.*

Shimizu, K., Fujita, R., Kondo, R. Sakai, K., and Kaneko, S. 2003. Morphological features and dietary functional components in fruit bodies of two strains of *Pholiota adipsa* grown on artificial beds. *J. Wood Sci.* 49:193–196.

Šišković, N., Strojnik, L., Grebenc, T., Vidrih, R., and Ogrinc, N. 2021. Differentiation between species and regional origin of fresh and freeze-dried truffles according to their volatile profiles. *Food Cont.* 123:107698.

Siwulski, M., Rzymski, P., Budka, A., Kalač, P., Budzyńska, S., Dawidowicz, L., Hajduk, E., Kozak, L., Budzulak, J., Sobieralski, K., Niedzielski, P. 2019. The effect of different substrates on the growth of six cultivated mushroom species and composition of macro and trace elements in their fruiting bodies. *Eur. Food Res. Technol.* 245:19–431.

Slutsky, A. 2021. Nametake Enoki Mushroom. *Food52*.

Smith, K. 2021. Ingredient spotlight: Mirin, a sweet umami rice wine that's an essential in Japanese cuisine. *One Green Planet*.

Soon, J. and Bong, K. 2013. A study of morphological characteristics and hybridization on *Lepista nuda*. *J. Mush.* 11:1-8.

Soulard, J. 2022. Béchamel sauce. *Mordu - Radio Canada*.

South Vancouver Island Mycological Society. 2022. Annotated checklist for larger fungi on Vancouver Island. *SVIMS*.

Spahr, D. 2024. (Untitled post) *Fungi* magazine (Facebook)

Specialty Produce. 2023. Himalyan black truffles. *Specialty Produce*.

Stamets, P. 2005. Mycelium running: How mushrooms can help save the world. Ten Speed Press, Toronto. 339 pp. Reviewed at *Amazon*.

Stott, K. 1998. Characteristics of Australian edible fungi in the genus *Lepista* and investigation into factors affecting cultivation. *Ph.D. Thesis Univ. West. Sydney Hawkesbury.* 167pp.

Svanberg, I. and Lindh, H. 2019. Mushroom hunting and consumption in twenty-first century post-industrial Sweden. *J. Ethnobiol. & Ethnomed.* 15:42.


Tacitus, Publius CorneliusTacitus, Publius Cornelius. ca. 116 A.D. *Annals*.

Taschen, E., Sauve, M., Vincent, B., Parladé, J., van Tuinen, D., Aumeeruddy-Thomas, Y., Assenat, B., Selosse, M., and Richard, F. 2020. Insight into the truffle brûlé: tripartite interactions between the black truffle (*Tuber melanosporum*), holm oak (*Quercus ilex*) and arbuscular mycorrhizal plants. *Plant Soil* 446:577–594.

Taste of Home. 2010. Canadian cheese soup. *Taste of Home*.

Tejedor-Calvo, E., Amara, K., Reis, F., Barros, L., Martins, A., Calhelha, R., Venturini, M., Blanco, D., Redondo, D., Marco, P., and Ferreira, I. 2021. Chemical composition and evaluation of antioxidant, antimicrobial and antiproliferative activities of *Tuber* and *Terfezia* truffles. *Food Res. Int.* 140:110071.

Tejedor-Calvo, E., Morales, D., Marco, P., Sánchez, S., Garcia-Barreda, S., Smiderle, F. R., Iacomini, M., Villalva, M., Santoyo, S., and Soler-Rivas, C. 2020. Screening of bioactive compounds in truffles and evaluation of pressurized liquid extractions (PLE) to obtain fractions with biological activities. *Food Res. Int.*, 132:109054.

Thongbai, B., Wittstein, K., Richter, C., Miller, S., Hyde, K., Thongklang, N., Klomklung, N., Chukeatirote, E., and Stadler, M. 2017. Successful cultivation of a valuable wild strain of *Lepista sordida* from Thailand. *Mycol. Prog.* 16:311–323.

Thu, Z., Myo, K., Aung, H., Clericuzio, M., Armijos, C., and Vidari, G. 2020. Bioactive phytochemical constituents of wild edible mushrooms from Southeast Asia. *Molecules* 25:1972.

Totally Swedish. 2024. Swedish elk meatballs with chanterelle sauce. *Totally Swedish*.

Toussant-Samat, M. 1987. History of Food (translated by Anthea Bell). Blackwell. 801pp. Reviewed at *Goodreads*.

Towns, K. 2024. Enjoy a cold weather feast with winter chanterelles. *Fungi (Epicure)* 16(5):31-34.

Trappe, J. 1962. Fungus associates of ectotropic mycorrhiza. *Bot. Rev.* 28:538-606.

Trappe, M. 2004. Habitat and host associations of *Craterellus tubaeformis* in northwestern Oregon, *Mycologia* 96:498-509.

TRUFF. 2024. Eggs Benedict with truffle Hollandaise. *TRUFF*.

Truffle Association of British Columbia. 2022. *B.C. Truffles*.

Tubić, J., Grujičić, D., Radovic-Jakovljevic, M. Ranković, B., Kosanic, M., Stanojkovic, T., Andrija, C., and Milošević-Đorđević, O. 2019. Investigation of biological activities and secondary metabolites of *Hydnum repandum* acetone extract. *Farmacia* 67:174-183.

Tung, C., Lin, C., Wang, H., Chen, S., Sheu, F., and Lu, T. 2018. Application of thermal stability difference to remove flammutoxin in fungal immunomodulatory protein, FIP-fve, extract from *Flammulina velutipes*. *J. Food & Drug Anal.* 26:1005-1014.


Uncut Recipes. 2021. Nametake recipe. *Uncut Recipes*.

United States Department of Agriculture, Agricultural Research Service. 2019. *FoodData Central*.

United States Department of Agriculture, Agricultural Research Service. 2022. *FoodData Central*.


Valsta, L., Kilkkinen, A., Mazur, W., Nurmi, T., Lampi, A., Ovaskainen, M., Adlekreutz, H., and Pietinen, P. 2003. Phyto-oestrogen database of foods and average intake in Finland. *Brit. J. Nut.* 89(S1):S31-S38.

Vanderbilt, T. 2016. The psychology of genre: What we don't like, we struggle to categorize. *The New York Times* May 28 2016.

Vasdekis, E., Karkabounas, A., Giannakopoulos, I. Savvas, D., and Lekka, M. 2018. Screening of mushrooms bioactivity: piceatannol was identified as a bioactive ingredient in the order Cantharellales. *Eur. Food Res. Technol.* 244:861–871.


Wang, N. 1995. Edible Fungi. Cyclopedia of China. Agriculture Printing House. Beijing, PR China.

Wang, P., Hsu, C., Tang, S., Huang, Y., Lin, J., Ko, J. 2004. Fungal immunomodulatory protein from *Flammulina velutipes* induces interferon-gamma production through p38 mitogen-activated protein kinase signaling pathway. *J Agric. Food Chem.* 52:2721-5.

Watanabe, A., Iwanaga, T., Tomita, T., and Shimizu, M. 2004. Effect of a pore-forming protein derived from *Flammulina velutipes* on the Caco-2 intestinal epithelial cell monolayer. *Biosci. Biotechnol. Biochem.* 68: 2230-2238.

WesO8647. 2011. Pittsburgh Potatoes. *Food.*.

Wikipedia. 2023. Cunda Kammāraputta. *Wikipedia*.

Wilson, C. and Trotter, C. 2005. Scottish traditional recipes—A heritage of food and cooking. Arness Publishing Ltd., London. 256 pp. Reviewed at *Amazon*.

Winder[1], D. 1871. The Mushrooms of Canada: With engravings, and catalogue of the fungi of Canada. Ministry of Agriculture, Toronto. 24pp. *Archive.org*.

Wiringa, J. 2023. Chicken roulade or ballotine. *Jules Cooking*.

Wood, M. and Stevens, M. 2021. California fungi—*Hydnum oregonense*. *Mykoweb*.

Wu, L., Gao, X., Duan, Y., Bian, Y., Yang, T. and Fu, M. 2017. Testing of antiviral characteristics of flammutoxin in transgenic tobacco. *J. Plant Dis. & Prot.* 124:429-435.

Wu, Z., Meenu, M., and Xu, B. 2021. Nutritional value and antioxidant activity of Chinese black truffle (*Tuber indicum*) grown in different geographical regions in China. *LWT* 135:110226.


Yang, W., Yu, J., Pei, F., Mariga, A., Ma, N., Fang, Y., and Hu, Q. 2016. Effect of hot air drying on volatile compounds of *Flammulina velutipes* detected by HS-SPME–GC–MS and electronic nose. *Food Chem.* 196:860-866.

Yang, X. 1986. Cultivation of edible mushroom in China. Agriculture Printing House, Beijing, PR China.

Yates, L. 2023. The best creamy salmon pasta. *Foxes Love Lemons*.

Yeh, M., Ko, W., Lin, and L. 2014. Hypolipidemic and antioxidant activity of enoki mushrooms (*Flammulina velutipes*), *BioMed Res.* 2014:352385.

Young, C. 2023. Freezing makes the perfect beef wellington foolproof & wasy. *Chris Young* Dec. 14, 2023.


Zhu, C., Li, Z., Li, D.and Xin, Y. 2014. Pb tolerance and bioaccumulation by the mycelia of *Flammulina velutipes* in artificial enrichment medium. *J. Microbiol.* 52:8–12.

Zhu, M., Zhang, G., Meng, L., Wang, H., Gao, K., and Ng, T. 2016. Purification and characterization of a white laccase with pronounced dye decolorizing ability and HIV-1 reverse transcriptase inhibitory activity from *Lepista nuda*. *Molecules*. 21:415.

---

[1]     NB: No direct relation to author

# Series Index

Matsutake **1:** 3 **3:** 2,4-6,22,79,84-99 **4:** 2,34,51
Mercury **1:** 38 **2:** 6,50,71 **3:** 5,26,28 **4:** 5
*Morchella* (see morels)
Morels **1:** 1-2,6,8,11,20,33-34,37-38,40,43-59 **2:** 1 **3:** 1,45,101 **4:** 1,18
Mushroom alcohol (octenol) **1:** 3,49-50,64,83,88 **2:** 12,15,50,79,84,89 **3:** 26,54,64,91,94-95,106 **4:** 11,18,29,51
Mycorrhiza **1:** 38,46-47,80 **2:** 6,48,57,69 **3:** 5,9,11,19,21-23,28,79,85-90,102-104 **4:** 5,10,15-16,27,42-44

Nickel **1:** 38 **2:** 6,71
Nitrogen **1:** 46,62-64,81 **3:** 43,53,71,90,103 **4:** 10,43,51
North American dishes **1:** 8,17-19,22-28,52-53,57-59,65,69,71,73,78,84,90 **2:** 18,20,22-23,26,29-33,38-39,46,52,54,66-67,72,76,100-102 **3:** 15-16,30,32,37-38,47-49,51,57-60,98,108 **4:** 13,24-25,53,57,60-61,68-69

Oceana dishes **1:** 17,29
Octenol (see Mushroom alcohol)
Onion and *Allium* spp. Dishes **1:** 7-11,13-14,16-19,21-29,32,53,57-59,65,65-74,76-78 **2:** 19-24,26-46,53,55,60-62,66,72,76,81,85,90-93,97,100-102 **3:** 16-17,30-33,36-38,40,47-51,57-60,67-77,82-83,97,99,109 **4:** 13,20,22-24,31-35,39-40,53,57-60,62-73
Oyster mushrooms **1:** 1,11,21-21,26,37-38,40,46,60-78 **2:** 79,81 **3:** 15,50,53 **4:** 34,68

Paddy straw mushroom **2:** 5-6,77-81
Pasta **1:** 66-68,74 **2:** 29-31,33-34,37,40-42,55,60-62,76,80-81,92 **3:** 33,36,67,69,73,75-76,83,109 **4:** 40,58-60,62,64-67
Pastry **1:** 54,75 **2:** 53,72,101 **3:** 39-40,50,60,77 **4:** 21,49,55,71-73
Pathogen **1:** 46,50,81 **2:** 57,101 **3:** 42-43,63,65,85-87 **4:** 11,42
*Phallus* spp. (see Bamboo mushroom)
Philippines (see Oceana dishes)
Phosphorus **1:** 37 **2:** 4-5,95 **3:** 4,80,88 **4:** 4,10,43
Pine mushroom (see Matsutake)
*Pleurotus* spp. (see Oyster mushrooms)
Poisons (see Toxins)
Polynesia (see Oceana dishes)
Pork and ham **1:** 18,67-68,74-76 **2:** 20,29-33,38,55,60-62,66,76 **3:** 16-17,77,98,109 **4:** 34,57,59-63,66-68
Portabellas (see *Agaricus*)
Potato dishes **1:** 18 (potato flakes),25 (serving suggestion),78 **2:** 26-27,43-44,66,72,85-86,92 **3:** 32,68,70 **4:** 23-25,57,68-70
Potassium **1:** 37 **2:** 4-5,84 **3:** 4,80 **4:** 4
Poultry (see Chicken & poultry)
Puffballs **2:** 68-72 **3:** 21

Rice dishes **2:** 21,27-28,35-36,45,52,81,91(vinegar),**3:** 16-17,31,37,77,97-99 **4:** 20,32-33
Roasted mushrooms **1:** 13
Rubidium **2:** 6,50,79

Saffron milk caps **3:** 3-6,8-17
Sandwich **1:** 24,56,69,73,76-77 **2:** 46,67,101-102 **3:** 38,48,75 **4:** 54
Saprobe **1:** 46 **2:** 69,74,83 **4:** 9-10,37
Sauces **1:** 20-32,78 (butter sauce) **2:** 21,24,26-27,29-31,33-37,40,-44,55,60-61,66,81,100-102 **3:** 33-37,39-40,47,49,51,59,73,76-77,98-99,109 **4:** 23-25,32-35,58,60-63,65-73,79-83
Sautéed mushrooms **1:** 3,5-11,15-19,26-28,35,51,53-55,57-58,66,70,72-73,75,78,90,100 **2:** 23,25,27-31,33-46,53-54,60-62,66,76,101-102 **3:** 12,15-17,27,30-31,33,35-40,47-51,57,59-60,64,67,69-70,74,76-77,82-83,93,98,109 **4:** 11,13,20-25,34-35,39,49,58,62-63,66,68-70
Seafood **1:** 20 (fish sauces),59,71,73 (oyster sauce),78 **2:** 45-46,80,85 (algae) **3:** 34-35 (fish sauce),51,57 (imitation),60 (imitation),68-69 (kelp/fish sauce),83 **4:** 65-66
Seasoned powder **1:** 3,5,13,17,21-22,25,27,31,66,71-72,77,90 **2:** 3,19,21,25,44,72 **3:** 2,15,38,48,58,69,77,98 **4:** 2-3,34,40,55,66,69
Selenium **3:** 4-5,26,28,64 **4:** 4
Shaggy manes **1:** 1,37-38,40,80,85-91

# Acknowledgments

I'm grateful to my family and friends for supporting this project. I'm especially thankful for the patient assistance of my spouse Christene and daughters Emily and Hannah for tasting the dishes, and helping with recipe development. I also owe sincere thanks to Kem Luther for his review and corrections focused on the taxonomic information in the series. I would also like to express my appreciation to Shannon Berch for contributing the foreword. Finally, special thanks to Sarah Faye, Josh Aron, and Emily for their kind contribution of blewits, winter chanterelles, and hedgehog mushrooms used to test recipes in this part of the series.

# About the author

Richard Winder lives with his family on the forest frontier of coastal British Columbia. He is a retired research scientist (forest biologist and ecologist) who still studies the Earth's forests and writes non-fiction (about mushrooms) and science fiction. He published *Stella*, a science fiction novel available in print and electronic formats. Richard has also published eight other short stories in three different anthologies. He is active in a local mushroom club (the South Vancouver Island Mycological Society), and enjoys teaching people about the secrets of mushrooms and the fungal kingdom.